JAGUAR
FROM THE SHOP FLOOR
Foleshill Road and Browns Lane – 1949 to 1978

Brian James Martin

For Margot

With thanks to my agent, John Starkey, for his wise guidance and hours of effort

Also from Veloce Publishing –

Biographies

A Chequered Life – Graham Warner and the Chequered Flag (Hesletine)
A Life Awheel – The 'auto' biography of W de Forte (Skelton)
Amédée Gordini ... a true racing legend (Smith)
André Lefebvre, and the cars he created at Voisin and Citroën (Beck)
Driven by Desire – The Desiré Wilson Story
First Principles – The Official Biography of Keith Duckworth (Burr)
Inspired to Design – F1 cars, Indycars & racing tyres: the autobiography of Nigel Bennett (Bennett)
John Chatham – 'Mr Big Healey' – The Official Biography (Burr)
Tony Robinson – The biography of a race mechanic (Wagstaff)
Virgil Exner – Visioneer: The Official Biography of Virgil M Exner Designer Extraordinaire (Grist)

Jaguar

Jaguar E-type Factory and Private Competition Cars (Griffiths)
Jaguar XJ 220 – The Inside Story (Moreton)
Jaguar XJ-S, The Book of the (Long)
Great Cars: Jaguar E-type (Thorley)
Great Cars: Jaguar Mark 1 & 2 (Thorley)

www.veloce.co.uk

First published in May 2018 by Veloce Publishing Limited, Veloce House, Parkway Farm Business Park, Middle Farm Way, Poundbury, Dorchester, DT1 3AR, England. Tel +44 (0)1305 260068 / Fax 01305 250479 / e-mail info@veloce.co.uk / web www.veloce.co.uk or www.velocebooks.com.
ISBN: 978-1-787112-79-7; UPC: 6-36847-01279-3.
© 2018 Brian James Martin and Veloce Publishing. All rights reserved. With the exception of quoting brief passages for the purpose of review, no part of this publication may be recorded, reproduced or transmitted by any means, including photocopying, without the written permission of Veloce Publishing Ltd. Throughout this book logos, model names and designations, etc, have been used for the purposes of identification, illustration and decoration. Such names are the property of the trademark holder as this is not an official publication. Readers with ideas for automotive books, or books on other transport or related hobby subjects, are invited to write to the editorial director of Veloce Publishing at the above address.
British Library Cataloguing in Publication Data – A catalogue record for this book is available from the British Library.
Typesetting, design and page make-up all by Veloce Publishing Ltd on Apple Mac. Printed in India by Replika Press.

JAGUAR
FROM THE SHOP FLOOR
Foleshill Road and Browns Lane – 1949 to 1978

Brian James Martin

Contents

Introduction 6

Chapter 1: Why Jaguar? 8

Chapter 2: 1949 14
I join Jaguar, the NUVB, and Labour

Chapter 3: 1949-1950 20
The XK120, assembly line chaos, and Gladys and Dorothy

Chapter 4: 1950-1951 31
Earls Court Motor Show, my first date, and the Mk VII

Chapter 5: 1951-1955 40
The RAF, towing aircraft, and night flying

Chapter 6: 1955 52
Jaguar at Browns Lane, 24 Hours of Le Mans, and my first car

Chapter 7: 1956 .... **64**
A royal visit, Jaguar retires from racing, and preparing for Earls Court

Chapter 8: 1957 .... **70**
The XKSS, the fire, and a new position

Chapter 9: 1957-1958 .... **79**
The experimental department, a Ford van, and rallying

Picture Gallery **94**

Chapter 10: 1958-1959 .... **101**
The Mk II in competition, E1A, and E2A

Chapter 11: 1960-1961 .... **110**
The Grand Prix, my first Jaguar, and Sandy Lane

Chapter 12: 1961-1962 .... **126**
The Mk X, more racing, and the early lightweights

Chapter 13: 1963-1965 .... **140**
The lightweights, the XJ range, and a big decision

Chapter 14: 1965-1972 .... **157**
New jobs, Forward Engineering, and coming full circle

Chapter 15: 1972-1973 .... **162**
Old friends, assembly lines, and Browns Lane transformed

Chapter 16: 1974-1978 .... **169**
The XJC, the XJS, and leaving Jaguar

Epilogue: My Jaguars **173**

Index **190**

Introduction

When, as a callow youth of sixteen, I walked up Swallow Road for the first time, I hadn't the remotest idea of what was to follow. Now, over sixty years later, I am sitting down to recall some of the experiences of my working life.

Jaguar, and the Jaguar marque, filled an enormous amount of those years, and to some extent continue to do so. During the time I spent with Jaguar, I went from working on the shop floor, firstly at Swallow Road and then at Browns Lane, to the experimental and competition departments, before finally landing in management.

I have served on the board of the Jaguar Drivers Club, was a founding member, and later the chairman, of the Jaguar Car Club, and am a member of the Jaguar Enthusiasts Club. After retirement, I spent five happy years as a volunteer with the Jaguar Daimler Heritage Trust, in the museum at Browns Lane. I then spent five years living in France, before returning home to England, and I still have a Jaguar: an XJ8 (X308), a magnificent touring car!

Over the years, I made many friends in the Jaguar world, but sadly so many of them are no longer with us, such is the penalty of survival. I have no great claim to fame, and could never be classified as anything other than a small cog in a much larger machine. However, I have lived through the making of the Jaguar marque, and played a part, no matter how small, in its history, and that is something I will always be proud of.

Me driving the ex-Ian Appleyard, Alpine Gold Cup winning XK120, accompanied by a Chinese journalist. (Author's collection)

I have in my possession a first-edition copy of Andrew Whyte's book *Jaguar: the Definitive History of a Great British Car*, in which Andrew wrote, 'to Brian, a true Jaguar enthusiast,' and that just about sums me up.

I have never kept a diary, so the stories are all from memory. I hope that you will enjoy these recollections, and will make allowance for my old age and reduced brain power. I trust that you, fellow enthusiasts, will find some interest in what follows.

Brian James Martin

Chapter 1

Why Jaguar?

I come from a very normal background. My father was a coal miner and my mother worked in service until she married. When I was six years old, Adolf Hitler made sure that my early youth would be a trial. Despite this and the effects of wartime England, including the rationing of just about everything, I had a happy, if somewhat sheltered, upbringing. My parents never owned a house, nor a motor car. To get to work, my father rode five miles on an ancient push-bike with acetylene lamps. The nearest bus stop was a two-mile walk and there were only two buses each day, Monday to Friday. I had the same two-mile walk to catch the school bus in the morning, and again in the evening to return home. The closest that we came to a motor car was the local taxi, although we could never afford to hire it. It was a 1934 Austin 16, but to me, it was a Rolls-Royce.

When I look back on my teenage years, I never fail to be amazed at how incredibly poor we were against today's standards. Keepers Cottage, Mancetter, was a typical late 1800s, two-up two-down, tied property, built, as the name suggests, to house the gamekeeper of the nearby Mancetter Mansion estate. The only running water we had was in the outside wash house, where the water was heated in a huge copper, fed by a coal fire, and carried to and from the house by bucket. Bath night was held in a large tin bath in front of the fire in the kitchen, with no other form of heating unless a fire was lit in every room. Winter time was akin to living in a refrigerator. My mother did our washing in the wash house using a zinc dolly tub, a scrubbing board, and a mangle with huge wooden rollers. Somehow the sheets were always snow white.

We had no electricity. Light either came from candles or paraffin lamps, and I had my own candlestick to take to bed with me. The sole contribution to modernity was an ancient radio, powered by a lead acid six-volt accumulator, which, once a week, I had to take to the nearest garage (three miles each way) to be charged. That became much easier when I received my first bike, though I hate to think what health and safety regulations would have made of a seven-year-old on an ancient bicycle with no brakes, carrying a glass container full of sulphuric acid, but I was young and innocent then!

Of course, with no running water or mains drainage, toilet facilities were, to say the least, a little crude. I hated that outside toilet. You could look up and see the sky through the tiles, and I always imagined there were nasty things in the bucket that might leap up and bite me. I can also think of much better uses for last week's *Daily Express*! Some good, however, did come out of it; we had marvellous vegetables, all fed by my father's diligent spreading of the bucket's contents, as a sort of compost.

Wartime was tough for everyone, including us children. Many of the little luxuries of life that are now taken for granted simply did not exist. It would be years before I saw a bar of chocolate, a banana, or an orange, but we survived and are no worse for it.

My father was a sergeant in the Home Guard (which was not at all like the television show, *Dad's Army*) and we displayed a sign in the front window advertising that fact; what the German invaders would have made of that, God only knows. Dad's personal weapon was a Lee Enfield .303 bolt action rifle, which he taught me to strip and clean. I spent many hours with that rifle, though I hasten to add that I was never allowed near the ammunition. Many years later, when I joined the Royal Air Force (RAF), my drill sergeant was amazed at my dexterity when it came to handling the same weapon, and I think that I gained a little respect when I explained how I had attained the knowledge. I became a marksman at Catterick Camp in the annual competition and I swear that most of it was down to those early years, knowing how to handle and look after the old .303.

As a child, my first love was aeroplanes. I wanted to be a pilot and fight 'the Hun' attacking my country, but I was born too late, and my father, who I know felt the same way, was born too early. I recall visiting Coventry, immediately after the terrible night in 1940 when the Luftwaffe targeted the city, and seeing the awful devastation of war, not knowing that when my school days came to an end I would spend most of my working life there.

At the age of eleven, I had to choose to take examinations for either the local grammar school or an engineering scholarship. Having always been interested in all things mechanical, and considering grammar school kids to be snobs, I chose the latter. Surprisingly, I passed with flying colours, and I spent the next five years as a day student at Coventry Junior Technical College. I have always been grateful for the introduction to engineering that this provided. I excelled at engineering drawing, workshop practice, woodwork, physics, chemistry, history, geography, and English. I was dismal at mathematics, languages, and athletics. I did enjoy ball games though, and I became captain of the school football team, wicket-keeper for the cricket team, and even tried my hand at rugby until a cracked collarbone told me that it was too dangerous a pastime.

Upon leaving college, my grades were not brilliant, but passable. My parents talked with me long and hard about my future. I still wanted to be a pilot and had been a member of the Air Training Corps (ATC) for several years. Our science teacher at Hartshill School was also commander of the local ATC squadron. He had been a pilot before the war and was a squadron leader in the Royal Air Force Volunteer Reserve (RAF/VR), proudly wearing his wings. He, above all, was responsible for maintaining my interest in aeroplanes, and I also attained my own wings, although only as a glider pilot.

Just three miles from our house, as the crow flies, was Lindley aerodrome, known as the home of the Motor Industry Research Association (MIRA). Originally built as one in a ring of defensive airfields around the industrial Midlands, it had now been taken over by the United States Air Force (USAF). It was operating as a heavy bomber station, and I would see B-17 Flying Fortresses set out to bomb the Third Reich on a daily basis. I have vivid memories of them returning in the afternoon, trailing smoke, props feathered, huge holes in the flying surfaces, and, only a few hundred feet above our house, on the glide path for the main runway. Years later I would be driving Jaguars over those very same runways and perimeter tracks, which were still in use as MIRA developed. It really is a small world.

However, the war was over and there was no call for heroes to save the country. I was told that my dismal mathematical abilities would let me down in the modern air force, so I decided I would be an engine driver. Just a short distance from our house was a bridge that overlooked the main London to Scotland LMS line, and I would sit there for hours to watch William Stanier's Pacifics thunder under me. I collected all the numbers of the Jubilee Class – with names such as, Swaziland, Mauritius, and the Leeward Islands and imagined what those places were like – as well as those named after the Royals: Princess Margaret Rose, King George VI, Queen Mary, and the rest.

Soon enough, my father convinced me there was no future in that either. The last thing that he wanted me to be was a coal miner, as he had been, so eventually we decided that my future would be in engineering, specifically an apprenticeship, but where?

My letter writing then was quite passable, so, with a certain confidence, I picked out the companies who might have the honour of employing me. I chose motor manufacturers because if I couldn't work with planes, motor cars seemed like the next best thing. Plus, motor cars had wheels and wheels fascinated me. I had gathered together a pitiful selection of pre-war copies of *Speed* magazine, before it was amalgamated with *Motor Sport*, and I knew a bit about Mercedes-Benz, Alfa Romeo, Bentley, Brooklands, and 24 Hours of Le Mans, but I had little to no knowledge of SS Cars, or Jaguar as it had now become.

My first letters went to Rolls-Royce at Crewe (where else?), Alvis, Riley, Aston Martin, and Standard Triumph. I waited with baited breath for the responses. When they came, the message was loud and clear. From Rolls-Royce, "thank you for your enquiry, but no, we have no vacancies at this time"; from Alvis, "you have insufficient qualifications"; from Riley, "we are not taking on any further apprentices but please ask us in 12 months time." Aston Martin would 'get back to me' and never did. Standard Triumph would let me know when an interview could be arranged. What was wrong with these people? There I was, offering my services and no-one was interested.

During this time, my brother, who was eight years older than me, was courting a girl whose father worked in the Jaguar body shop at Swallow Road, having followed the company down from his home in Blackpool. He suggested that I should write to William Lyons, the owner of Jaguar, and request an interview. I took him at his word and did just that.

Several days later, I had a response in the post from a lady called Alice Fenton, saying that she had arranged an appointment for me at the employment office in Swallow Road. She also wrote that their apprenticeship quota was already filled, but that I might be considered for normal employment, pending the reopening of the apprenticeship scheme.

Well, this was tantamount to the offer of a job, as by now I had spent my first week's wages, was considering which motor car I would buy, and how I would respond when they offered me the post of chief engineer, such is the temerity of youth.

The interview was set for April 5, 1949, and my school term would finish soon after. It wasn't long before this that I had acquired my first set of wheels. I had scraped and saved to buy a second-hand Sunbeam lightweight cycle frame, in the must-have Reynolds 531 tubing size. A set of Derailleur gears, lightweight brake calipers, and Dunlop racing tyres on alloy rims saw me ready for anything that the world could throw at me. This bike was to be my only means of transport for the foreseeable future, and would also introduce me to the joys of a cycling club membership, as well as the opposite sex.

In an exceptionally bold move, when my father offered moral support, I refused, deciding to attend the interview alone. This all seems very farcical now, but at the time, I was making decisions that would change my world forever.

On the morning in question, dressed in my first long trousered suit, complete with cycle clips, leather gloves, and white socks, I set off on the 29-mile ride to Coventry. I was not to know that I would be repeating this ride for many years to come.

Old age has inevitably withered plenty of memories, but turning into Swallow Road for the first time is not one of them. I remember riding through the steel gates with the Jaguar logo emblazoned on them, past the Dunlop factory on my left, and on to the gatehouse, where I was directed to the employment office, just inside the gates.

I was among some twelve or so other young souls waiting for the same interview. It felt like forever before my name was called, and I was introduced to a rather rotund gentleman sitting at, what was to me, a table of the inquisition. I must admit that I do not recall his name, but he asked questions about my schooling, my pastimes, my interests, my family, before the door burst open and in walked a tall man with immaculate grey hair, steel framed glasses, wearing an impeccable striped suit and all that goes with that. The rotund gentleman then leapt to his feet. The man in the suit, ignoring my interviewer, looked straight at me and said "Well, are you here for a job?" "Yes," I

My second-hand Sunbeam bicycle outside our house. (Author's collection)

stammered. "Good," he replied. "If you get it remember that we only pay for results." With that he turned and walked out, and the rotund gentleman subsided into his chair, waving me away. I'd later learn that I had just met William Lyons!

The ride home seemed to take twice as long. My mother and father wanted to know what had happened, down to the very last detail. "That's that then," was my father's response, whilst my mother made sympathetic noises. A few days later a letter arrived in the post from none other than Alice Fenton. She had been instructed to advise me that there was an open position in vehicle assembly, which I could take up with a view to transferring to the apprentice scheme as and when a vacancy became available.

To this day, I have no idea why Alice Fenton was so involved in my employment. It was certainly not normal practice, as such correspondence would usually have come from the employment office. Many years later, after Sir William's retirement, I happened to introduce him at a club function, and during conversation I could not resist the temptation to ask him about my experience. He looked very hard at me and asked, "Were you a good employee of Jaguar?" to which I responded, "I would like to think so." He said, "Well, we were right to hire you then, were we not?"

And so, in mid-April, 1949, I cycled up Swallow Road, past the Dunlop factory, and through the steel gates, to start my life at Jaguar.

What follows is a window into that life.

Chapter 2

1949

I join Jaguar, the NUVB, and Labour

On the morning when I officially became part of Jaguar's workforce, I reported, as instructed, to the employment office. After the formalities of registration, etc, I was told to report to Mr Lee in the production offices. I pointed out that I did not have a clue where these were, and was given directions, which seemed to be a test of initiative.

As I walked up Swallow Road, I observed the nature of Jaguar's facilities. The building on my immediate right – at the north end of the street, where the ground sloped down to the boundary with the Dunlop factory – was a very old, rambling, and somewhat dilapidated, wooden structure built on stilts. I would soon learn that this contained the body shop, white metal department, experimental department, maintenance, wood shop, the final production line, and despatch department.

This timber-framed building had an illustrious history. During the First World War, it was owned by ammunitions producing company White and Poppe, and was used as a shell filling factory, which must have been a pretty dangerous place to work. William Lyons had leased this factory when he moved production from the Blackpool site, where the then named Swallow Company had started.

I am amazed that, even as I write today, this building still exists, although heavily updated. It is not generally known, but during the First World War, this part of Foleshill, near Coventry, became one great industrial site. The Daimler factory on Sandy Lane was at the heart of it all. Daimler made an enormous contribution

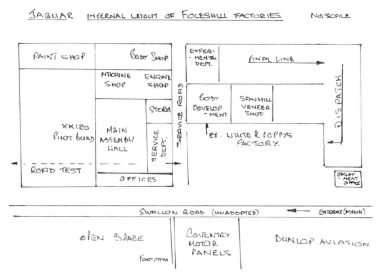

A schematic layout of the Foleshill Road Jaguar factory, circa 1949. (Author's drawing)

to the war effort, building staff cars, armoured cars, and support vehicles, which were all built at the Radford works. Not only this but complete RE8s and SE 5A aeroplanes were built and flown from Radford aerodrome, adjacent to the works. A complete village was built around the White and Poppe's shell filling complex, with its own living quarters, shops, canteens, and even a cinema.

Separated from this old building by a road, was a much newer building, leased from Dunlop's Aviation Division. This housed the main offices, the paint shop, vehicle assembly, the trim shop, the machine shop, the service department, and the road test department. This building had been leased to facilitate Jaguar's work on military projects during the war, and was now given over to vehicle production.

The Jaguar production offices were situated in this new building, adjacent to the assembly lines. I eventually found an office which had a painted notice on the door saying 'production control,' and I timidly knocked on the door. It was opened a few seconds later by a very pleasant man who, when I enquired about the whereabouts of Mr Lee, explained that he was on the shop floor but was expected back soon, and I could wait for him in the office. He obviously realised that I was very nervous and did his best to put me at ease. He explained that the production control office was the seat of Jaguar's empire, housing the foremen who oversaw all of the vehicle assembly production lines, chassis build, pre body mount, mounting, and trim tracks.

Just as he was really getting into it, the door opened, and in walked a tall, sombre-looking man wearing a white cow gown with a blue collar. This was George Lee.

George, who I came to know as a friend after his retirement, had moved with the company from Blackpool, and was an integral part of the system. He never aspired to high status, but was hugely respected for his abilities to lead the workers and get the job done. He came across as a tyrant, but at times, his Lancastrian wit could be very funny. I later found that he was a highly intelligent man, whose bark was much worse than his bite, although he did not suffer fools lightly.

"Well lad, what can I do for you?" he asked. I explained that it was my first day at Jaguar and had been told to report to him. "Alright," he said, "let's see if we can find you a job that won't give us a problem. Follow me." In a flash, he disappeared back through the door, with me in hot pursuit. We wound our way over the assembly lines full of new Mk Vs, gleaming paintwork, noise, fumes, and apparent mayhem. We finally came to a stop where the painted bodies left the paint shop. George called over a young man wearing a brown cow gown with a green collar. "This is your leading hand," said George. "He will show you where you will be working and arrange for everything that you need." I had just met Peter Craig.

Peter was a canny Scot who eventually became a director and a highly respected member of senior management. I'm proud to say that he became a firm friend who, despite his position and the fact that I was still wearing overalls, would always stop to talk to me, particularly about old times. I must admit this caused some consternation among my workmates, who I'm sure believed me to be a management spy.

Peter completed the necessary paperwork for me to draw my basic tools and overalls from the production stores. The tools were a ball-peen hammer, a large

The Mk V: this one a drophead coupé. (Courtesy John Starkey/ Jaguar Daimler Heritage Trust)

screwdriver, a pair of pliers, an adjustable spanner, a rubber mallet and a large bastard file. I was told that if any special tools were required I could sign for them at the stores, and would be responsible for their safekeeping and eventual return. Any other tools I needed, I would have to buy on the company purchase scheme, and the money would be deducted from my salary over a number of weeks. The overalls that I was supplied with were brown and three sizes too big. To this day, I am convinced that it was a deliberate part of some strategy for new employees.

Someone must have decided that I had an aptitude for electrical wiring, God knows why, and I was placed with an operator named Doug, who would show me what to do. My first operation was on the Lucas voltage control unit on the bulkhead of a Mk V body. I had to trim the wires to length, bare the end of the cables, connect them to the control unit, then wire up the fuse box, and fit the two-pin auxiliary plug. This all seemed very easy until I realised that I had to do this in half the time that I was taking, and on every single body that came to my station until the end of the shift.

Towards the end of my first working day, Peter came around and asked me where I lived and how was I getting home. He seemed surprised when I told him that I would be riding my bike, which I had left behind the gatehouse, to the other side of Nuneaton. "Tomorrow, you can park your bike in the cycle rack outside the offices, and avoid the long walk up Swallow Road" he said. This was to have substantial implications, which I will explain later.

I remember arriving home that evening, tired, dirty, and still wearing my overalls. I think that I would have slept in them had my mother allowed it. I had become a member of a select society; we made motor cars!

Peter Craig, who had now taken me under his wing, told me that he would move me around the workplace so that I could gain more experience. True to his word, I soon found myself fitting the chrome door frames, glass, and window winding mechanism on the nearside of the Mk V body, whilst another operator completed the offside.

Shortly after I started at Jaguar, I had my first run-in with the shop steward. He was a little weasel of a man with shifty eyes, and I took an instant dislike to him. "You will have to join the union to continue working here," he stated with some pomposity. At the time, the only union that I knew of was definitely of a sexual nature, so my immediate response was that I would think about it and talk to my parents. He went off in high dudgeon, telling me that he would return with the necessary forms to allow the membership subscription to be deducted from my salary, and that I would also be expected to be a card-carrying member of the Labour party. Panic!

I spoke to Peter Craig about this and he told me that unfortunately the company had an agreement with the unions and that, although it wasn't mandatory, I would be wise to join. Therefore, with some reluctance, a few days later, I became a full member of the National Union of Vehicle Builders (NUVB). My father was in full agreement, and suggested that I should also join the Labour party; he had always voted for them and thought I should do the same. By now, I was beginning to get to grips with what went on in the world, and I had an immediate distaste for anyone who told me what I should do. My father and I had our first political fall out before he relented, on the understanding that I would never vote Conservative. I never did tell him the truth!

This was a particularly bad time for industrial relations in Britain. The trade unions were beginning to flex their muscles, and they were highly politically motivated. The real problem for Jaguar, like most other manufacturers in Britain, was that there were just too many of them. All skilled trades were represented separately. For instance, most assembly line workers were members of the NUVB, body assembly personnel by the Tinsmiths, semi-skilled and unskilled workers by the Transport and General Workers (T&GW), trim and woodwork labourers by the Coachbuilders, to name a few. Not only that, if an employee was moved between departments as regularly happened, he did not have to change his union membership. The result was that an assembly line's labour force could have as many as three or four different unions represented.

The dreaded word 'demarcation' appeared, which promoted infighting between various factions. You could have two people working side by side, one a member of the NUVB, the other of the T&GW. If a dispute occurred elsewhere in the plant involving the withdrawal of labour by the T&GW, that worker would be expected to walk out with them, which resulted in chaos. However, a small pool of labour did exist, mainly to cover absenteeism, and would be allocated at the beginning of each shift.

All shop stewards had a particular operation to carry out, but in reality, they could rarely be found doing it as there was always some 'official union business' they had to attend to. Mr Shifty Eyes was very clever at using this as an excuse to skive off. I recall one occasion when he was confronted by the line foreman and asked to explain his absence. The reply was, "You're management and I am not at liberty to discuss the matter with you."

It was a few weeks in, and I had already made a few friends of a similar age. One of our pastimes during the lunch break would be to play football on the piece of waste ground north of the main building. The ball was normally a large lump of dried rubber solution from the trim line. This bounced very well but was quite painful if you were unfortunate enough to get in its way. When it was wet, we played the game in the men's toilet block, which was a peculiar affair mounted on stilts over the pre-mount line and accessed by a steel staircase. We were playing up there one day (during work time) when a shout went up: "Quick, the old man's coming." With that we all raced for the toilets, shut the doors and crouched on the seats. I heard the measured tread of feet, which stopped before seeming to recede. I thought that whoever it was had gone. Quietly climbing down, I crouched to look under the door only to find myself looking directly at a florid complexion, slicked back grey hair, and steel rimmed spectacles. The face quickly disappeared and the footsteps went away. In our panic, we had forgotten the ball which, when we emerged, had also disappeared. Fearing the worst, we all guiltily trooped down the stairs and went back to our work stations.

Sure enough, within a few minutes, I saw George Lee approaching. "Martin, I want to see you in the office. I think you know why."

Reluctantly, I dragged myself to his office in production control. I found two of my associates already there. On the table was our ball. To say that we were torn off a strip would not be an exaggeration. George spoke to us for five minutes, throwing distinct doubt on our parentage in the process. Bill Lyons had done the same to him, with emphasis on wasting company time and misuse of company property. Needless to say, we never played football in the toilet block again.

Chapter 3

1949-1950

The XK120, assembly line chaos, and Gladys and Dorothy

It was around August 1949 that I saw my first XK120. I was walking back up Swallow Road at lunchtime, after visiting the small café in Holbrooks Lane for tea and a sandwich, when I heard behind me the noise of an unfamiliar but powerful engine. I turned and saw a wonderful white sports car tearing up the drive. It was being driven by a lean man in white overalls and a white leather helmet, with a cigarette nonchalantly poised in his mouth.

This was HKV 500, the Jabbeke record breaker, where it recorded over 132mph with Ron 'Soapy' Sutton at the wheel. This must have been after the record run as the car sported a full width windscreen and was right-hand drive, which is what it was converted to after the run.

I was mesmerised. One day, I thought, I am going to have one of those, and I am happy to say that came true.

Of course, I knew about the new sports car, having read the report of the 1948 Earls Court Motor Show in *Motor* magazine, and understood the amazing level of interest that its announcement had created. But this was the first one that I had seen 'in the flesh.'

Some weeks later, Peter came to me and said "Collect your tools and follow me, you have a new job." He led me down the tracks, to an archway in the wall that separated the assembly hall from the north bay, which at the time was practically empty, and only housed the road test department.

Imagine my wonder when, in front of me, I beheld not one, but six XK120 bodies, mounted on mobile paint shop trolleys, all in different colours: red, dark blue, white, silver, light blue, and British racing green. Peter introduced me to a huge Scotsman called Les, who I recognised as an operator from the Mk V trim track. There were about half a dozen other operators present, all of long service, and well respected by those around them. Les explained that these first six cars would be hand built, as the production facilities were not yet ready to take them into main assembly (nor would they be for some time). I cannot remember how many cars we built this way, but it was far more than was expected.

I was to be the 'boy gopher,' the 'fetch and carry lad,' which would lead me into some hilarious (though not at the time) situations. One such situation occurred when I was told to go to the sawmill and bring back two sheets of plywood. I was to ask for a Mr Gardner and he would sort me out.

As it was raining heavily at the time, I thought it very considerate of Les to advise me to walk through the factory, then over the road to the old building where the sawmill was situated. I had negotiated the assembly lines, and the machine shop, and was halfway across what turned out to be the service department (a

The XK120 production line at Foleshill. A very early XK120 chassis (note the cast aluminium fan on the front of the engine), is being prepared to have the body mounted.
(Courtesy Jaguar Daimler Heritage Trust)

place I had never been before), admiring the rows of customer cars in for service and repair, when I was stopped dead by a loud shout. "Boy, where do you think you are going?" I was confronted by a well-dressed man who was at least seven feet tall, with bushy eyebrows and an angry face. Stooping down he said, "Well?" to which I stammeringly explained the nature of my journey.

At this, he went into even more of a rage, telling me that his department was not to be used as a short cut to anywhere, and that I would have to get permission to ever be there again. Watching this commotion, was a very military-looking man dressed in a dark uniform with shiny buttons, and a peaked cap. He was waved over and told to escort me back through the premises, which he duly did, with a friendly smile and a squeeze of my arm.

Later in life, during my Motor Club days, I would make a more friendly acquaintance with 'Lofty' England, despite our initial confrontation, but back then I felt uncomfortable in his presence.

Having been escorted out, I continued down the outside of the main building in the rain, entered the old factory, and promptly lost my way. I found myself in a large open workshop, in the middle of which was the unpainted body of a large saloon car. It had smooth flowing lines quite unlike anything that I had seen before. Several men were working on and around it. One of them stopped and asked me what I was doing there. I explained my mission and he pointed me in the direction of the sawmill. He also told me that I should leave at once before the 'gaffer' saw me. It was only then that I realised I had passed a sign saying 'Experimental body shop. Strictly no admittance.' I had stumbled into the build shop of the early Mk VII prototype. I hurried away in the direction indicated, with my mind in a whirl.

Eventually, I came to a sliding door marked 'sawmill and polishing shop.' Opening this door, I saw several ladies in the process of hand polishing the veneered sections of Mk V woodwork. I had to run the gauntlet of lascivious smiles and lewd remarks to get to Fred Gardner's office. These ladies were something else, with a reputation I later came to appreciate. My attention was drawn from the ladies by the arrival of a brown trilby hat, which hit me on the chest and fluttered to the floor. "Well, what do you want?" shouted an angry voice. This was definitely my day for angry people. Fred had a voice rather like sandpaper on the wood that his department relied on. I explained that I had been sent to collect two sheets of plywood, and presented the

A typical club racer of the 1960s. Anthony Archer's well raced 1950 XK120 at Silverstone. (Author's collection)

An aluminium bodied XK120 OTS (a roadster) in the USA. (Courtesy John Starkey)

requisition that had been firmly clutched in my hand through the whole of this saga. "Who sent you?" he asked. "Peter Craig," I replied. "It is needed for the new sports car."

This seemed to mollify him somewhat and he steered me into the wood store, where I was shown, asked to sign for, and removed the items in question.

Now, if you have ever tried to carry a sheet of 8ft x 4ft x ⅛in plywood in a high wind, you will understand my next dilemma. And I had two of them. It was still raining cats and dogs and I quickly worked out that forming them into a bow over my head was the easiest way of carrying them, while also protecting me from the rain. I staggered back into the top shop exhausted but triumphant, only to be greeted with howls of derision. "Why have you taken so long?" came from Les. When I tried to explain the saga, again, there were fits of laughter. "I knew Lofty would get him, it never fails," someone said. "How did you get on with Fred Gardner? He eats youngsters for breakfast," another called.

The reason for the whole situation was that the plywood was required to make the door casings, as the manufacturing system was light years away. All that Les had was a paper pattern with the rough layout. From that, he cut out the casing by hand, oversized, and then trimmed it to fit the inside of the door. Each one had to be produced individually as no two doors were quite the same. An allowance also had to be made for the thickness of the trim leather.

After cutting out the space for the door pocket, the plywood panel was then labelled with the body number and LH or RH designation. I then had to take the completed door casings to the trim shop for covering. It was there that I met Vicki, a slim, blonde, Polish girl, with very large blue eyes. It was love at first sight! From then on, I was always happy to visit the trim shop.

While I always enjoyed a trip to the trim shop, there were serious matters at hand; Jaguar had totally underestimated the impact that the XK120 would have on potential customers. It was only ever intended as a short run model, about 200 cars, to introduce its larger and more important big brother: the Mk VII saloon and its new twin overhead camshaft XK engine. Orders had flooded in, particularly after the XK120's debut in America. After the deprivation and hardships of the war years, everyone who had the money wanted one tomorrow. The XK120 was seen as the future of the production line at Jaguar, a quantum leap over what had gone before, with the performance only previously achieved by far more exotic cars, and a price and glamour that could not be surpassed by any other manufacturers of the time. Overnight, the competition became out of date, overpriced, and out of fashion.

The race was on, and William Lyons would spend many hours prowling around these early cars, always asking when they would be ready. The answer was never what he wanted to hear. The XK120 was nowhere near ready for production, even on a limited scale. The aluminium body alone took so long to build that the decision to change to steel was obvious; fitting the bought out steel pressings together was far faster than building the aluminium panels, which, at the time, were hand formed over wooden frames. On top of this, the ordering of other bought out components (items bought from outside sources instead of manufactured in-house) was in a pitiful state. On the first cars, Dunlop was due to supply the rubber 'D' section, which formed the basis of the crash roll pad around the cockpit and on top of the doors. In reality, it was not yet available.

This is where my large bastard file became useful.

A large roll of fabric covered rubber pipe, of almost the right diameter was delivered to the shop. I had to cut this to a rough length to suit the door tops, and the front and rear crash roll pads. I then had to file down the external canvas to the rubber, so that nothing showed through the leather trim covering. I spent hours doing this and had the blisters to prove it.

Every windscreen upright had to be tailored to the body in question by filing the holes in upright's aluminium casting until the fixing bolts would line up, at just the right angle, with the cage nuts in the body.

Correctly shaped door seals were also in short supply, and they had to be made up from sections of rubber, glued together with rubber solution. The tooling for the plywood floors was not yet ready, and they, like the door casings, had to be cut by hand. The main instrument panel, which was intended to be an aluminium pressing, also had to be hand formed.

It was a desperate situation which would not be relieved for some time.

An early production Mk VII Jaguar. (Author's collection)

Jaguar's lessons were learned. At least until the Leyland era, when it would find itself back in the same position. This instigated the start of what would later become pre-production build, where the problems of transferring engineering design to production capability would be ironed out, and advanced ordering of bought out items would be essential – in theory anyway.

Despite all the trials of production, work was in progress to create the XK120 assembly line in the main assembly hall. This required a great deal of reorganisation, and some reduction in the Mk V production capacity, in order to give the XK120 the space it needed. Eventually, the problems were sorted and bodies began to trickle out of the paint shop and on to this new line, which for the time being was still static.

Now the chassis assembly line had to contend with the production of two different chassis. Visually, the XK120 was not that much different to the Mk V, and on several occasions in the early days, when the tracks were not properly coordinated, much head scratching would go on when trying to fit a Mk V body to an XK120 chassis and vice versa.

I was transferred to the new assembly line with an ego boost; because of my experience I was made a rectifier/trainer, which meant I was to show others how to build the car. Peter Craig was made foreman and his coat went from brown to white, with a blue collar to match.

I had, during my stay in the north shop, become familiar with several of the drivers working alongside us in road test. One of these was Bertie Farndon. Bertie lived in Hinckley, and drove, as his everyday car, a DKW 1000. He was a quick driver, and his DKW two-stroke had the performance to match his capabilities. One day, he suggested that instead of cycling to work, he would arrange to pick me up and take me to work, which was a relief, to say the least. I took him up on his offer and we became firm friends in the process. Sadly, after I moved from production, and attained my own means of transport, I lost contact with him, although we would still cross each other's paths occasionally and reminisce.

I still cycled when the weather was good, parking my bike outside the offices as Peter Craig had suggested, which didn't turn out to be such a wise idea. It was during this time that the lads in the white metal shop played a well organised prank on me. Each evening I was finding it more and more tiring to cycle home. I initially put this down to my reduction in physical capability until one day, as I was riding along, my saddle bag parted company with the saddle on one side, quickly followed by the other. The whole thing fell to the road with a thump and a clang and suddenly the bike felt much lighter. On investigation, I discovered several pounds of lead loading stick, carefully taped together and secured in the bag. Not only this, but on further investigation, I removed the saddle and found that the hollow sections of the frame had been stuffed with the same lead sticks. God knows how long I had been riding around like this, but I had a good idea who the culprits were and was determined to get my own back when the time was right.

Ever since my first day at Jaguar, I had been making enquiries at the apprentice school; I still desperately wanted to wear those green overalls with the Jaguar logo on the pocket. It was to no avail, as there was still no space for me, but I would be informed when it was possible for me to take up an apprenticeship. Much later in life, Peter Craig told me that he thought my application had been blocked by George Lee, who considered me too much of a valuable commodity for his

A typical early XK120, as originally built in 1951, with steel wheels and spats. The leaping Jaguar is not original. (Courtesy John Starkey)

department to lose. I never knew the truth of this, which, I suppose, came across as a backhanded compliment at the time. However, I have often wondered how my life would have progressed had this not happened. I suspect that it would not have been much different; I would still have sought a position at Jaguar at the end of my apprenticeship, probably in engineering development, which is where I eventually ended up anyway, just by a different and more arduous route.

Slowly, the problems with the production of the XK120 were sorted, and at last the pre-mount line became mobile and was phased in with the chassis and mounting lines. It was odd to see the mounting track with five or six Mk Vs followed by a lone 120. Jaguar had never before produced volume builds on a mixed line, particularly with a different chassis (although the chassis were very similar). It was the production of the E-Type that would bring a dedicated sports car assembly line, but until then, all of the XKs were mixed in with the saloons.

Whilst Jaguar continued to prepare a production line for the XK120, in August, 1949, I went to my first motor race. Bertie had told me that the company had entered three XK120s in the Production Car Race at Silverstone. He asked if I would like to go with him, and of course I said yes. I was thrilled to see the cars that I was currently working on, not just represented, but winning. Proud of what we had accomplished, I had a lump in my throat for days after, and fixed a small Union Jack and a Jaguar badge to the handlebars of my bike to celebrate.

Bertie was very enthusiastic about the XK120, which he occasionally now had to test drive. He was, however, very critical about the brakes. "Believe me," he said. "We will have real problems in that department when the car gets out in the real world." He was right, of course, as brake fade when driven hard was always a frightening experience on the early cars. With a flash of inspiration (or inside knowledge), he told me that a new system would be needed to replace the old drums, something like that used on aeroplanes. Dunlop and Jaguar were of course already working on such a system, derived from the existing knowledge of Dunlop's Aviation Division. The disc brake was not far in the future.

By September 1949, the XK was starting to be produced in limited quantities, mostly for the export market, which was mainly North America. Very rarely at that time did one see a right-hand drive car on the assembly line.

There is a period factory photograph, that has been used many times, of a line of the cars in Swallow Road awaiting despatch to the States. This was not a regular occurrence, and was set up deliberately by Jaguar's publicity department to advertise what Jaguar was doing for the export market. American cars were despatched minus their headlights, the aperture being blanked off with a cardboard disc. At one stage, we also had a problem with tyre supplies, and Fred Gardner's shop came up with the ingenious method of making a wooden replacement from interlocking blocks. Obviously the car could not be driven on these, so slave wheels were used for road testing, before being replaced with the wooden variety prior to despatch.

As the end of 1949 approached, production was up to about 20 vehicles per day and some home market deliveries were being made, presumably to special customers. Most of the few right-hand drive cars that were produced were also being exported, mainly to Australia.

At about this time, Bertie's DKW developed a major engine problem and was off the road for about a month, whilst I was becoming less enthusiastic about cycling 40 miles a day, so I started getting the train. I was quite used to catching the train as the 'Tech' (technical college) I had attended was only a short walk from Coundon station. I still had a five-mile hike to get to and from Nuneaton, where I could catch the local train to Foleshill, and from there it was only a five-minute walk to the factory.

These were long, hard days, particularly in the winter. It meant leaving home at 6.30am to catch the train which arrived at Foleshill at about 7.45am, just about right for the 8.00am start. In the evening, the 5.30pm train would get me home at about 7.00pm. However, trains were rarely on time, and there was very often a mad dash up Swallow Road to clock-in by 8.03am. After that, a quarter of an hour was deducted from your pay. The only way to avoid this was to have your clocking-in card countersigned by your foreman, to say that you were ready to start work, and that was only allowed twice in any one week.

With the XKs now being produced with few problems, the work became monotonous, until one day in early December. We came to work one morning to find absolute chaos at the end of our track. The chassis assembly line, which ran alongside the line where I worked, had somehow switched itself on during the night. A number of chassis, with nowhere else to go, had climbed over one another, creating a jumble of upturned trolleys and bent link bars. To make matters worse, one of them had speared into the lead car on our line, badly damaging the nearside of the body.

The maintenance foreman was apoplectic, suggesting deliberate sabotage whilst his gang sorted out the mess. The mounting track had to stop as it had used the last chassis, and the electric hoist, which lifted the body from the pre-mount and lowered it on to the chassis, ran out of track after three car spaces. While we waited for our track to start, we stripped the damaged body, and returned it to the paint shop for repair and repaint.

Eventually, all was sorted, and production got under way, but we had lost about twelve cars in the process, which we could ill afford to do.

Christmas came and went, and a new year (1950), and a new era for Jaguar, began. Each day, I was getting bored of doing the same job over and over again, and I was relieved when someone was ill or absent because it meant I could take over a different operation, relieving me from the repetition.

We had two girls, Gladys and Dorothy, working on the line with us, who wired up the centre instrument panel. They built the panels themselves, assembling them trackside, and, even now, if you look behind your XK instrument panel, you could find the name Gladys or Dorothy, and the chassis number. They would assemble the panels trackside and then sit on a small upholstered seat over the gearbox tunnel, with the completed panel balanced on their laps, and attach the wiring, oil pressure pipe, revolution counter, and speedo cables, before fixing the panel in place. It can be seen that anyone working on the inside of the engine bay could peer through the gearbox tunnel aperture and hope to see a flash of thigh. The girls, however, were well aware of this. Dorothy would always wear slacks and Gladys had a neat way of folding her legs underneath her. General disappointment all round!

Trim components were delivered to the track on a continuous overhead conveyor. The outer panel, at that stage only designated by trim colour, was then built up by the girls using the production schedule, which was updated every day. This schedule carried all of the relevant information, such as date of order, colour, trim colour, axle ratio (for the correct speedometer), distributor or customer, LH or RH drive, and any special requirements, and was used through the whole production build of the vehicle, a ticked off copy travelling with each car.

At the start of the new year, Gladys left to have a baby, and I was appointed to carry out her operation whilst training a temporary replacement. It was only then that I realised how nimble-fingered and quick the girls were. I really struggled to keep up with Dorothy and I was exhausted at the end of the first day. Pauline, who I was training, took to it like a duck to water, and after just a few days was far quicker than me, particularly in the car; I had a new-found respect for those girls.

The XK120 was already famous, so we had several visits from important people. The young Duke of Kent and his entourage toured the factory, and a forever smiling Clark Gable, who was a very early XK customer, made the girls in the trim shop swoon. When touring the line, his knowledge and intelligence were obvious, asking all the right, and sometimes critical, questions. Tony Curtis and George Formby were also visitors. George, himself a Lancastrian, insisted on being photographed with George Lee, alongside a Mk V drophead; they were a truly comical couple.

Months passed, and soon it was time for the 1950 Earls Court Motor Show. Near the date, October 27, and at very short notice, we were told that the company would be setting up a free trip for all employees who wanted to go and see the company display. It would include the train journey, the entrance ticket, and 10 shillings to spend. They told us this would all depend on there being enough people to fill the train, but as it turned out, so many employees applied that an additional train was required. Not only did I decide to go, I decided that now was the time to make a move on Vicki from the trim shop.

Chapter 4

1950–1951

Earls Court Motor Show, my first date, and the Mk VII

My next visit to the trim shop was solely for the purpose of talking to Vicki. I had absolutely no legitimate reason for being there, and what's more, I had not reckoned with Percy Leeson.

Percy was the superintendent of the trim shop, which was set in a totally enclosed area in the centre of the production lines, separated from them by an eight-foot, one-third glazed, steel partition. His office was in one corner of his little empire. He was a diminutive, dapper little man who had a very abrasive personality and a voice to match. I never did understand why Jaguar seemed to attract people of this nature in its junior and middle management positions; people whose whole purpose in life seemed to be belittling those they considered to be lesser mortals, and doing their best to make life difficult for us. Little wonder that industrial relations were at such a low ebb.

Anyway, when he accosted me at the entrance, Percy had every right to query why I was in his department. As I entered, he happened to be bending down to look at the back of a half trimmed seat, and I had not seen him. "Well, boy, what do you want?" was his opening remark. Now, I could hardly try to explain to him that I was hoping to talk to one of his girls, especially when she would be working, so, with a flash of inspiration, I came back with, "Mr Leeson, I understand that you want someone to clean your car in the lunch break." I should hasten to add that this, by the way, was not just lucky guess work on my part.

The Mk VII saloon, introduced in 1951, with the world renowned XK 3.4-litre straight-six OHC engine. (Courtesy John Starkey/Jaguar Daimler Heritage Trust)

Percy was one of the honoured few of middle management who were allowed a vehicle supplied by the company: a 1.5-litre saloon, later known as the Mk IV. Part of the arrangement was that everything after that, even if carried out by the company, was at Percy's expense. Percy and his money were not easily separated, so once a week he would arrange for a friend of mine, Ralph, who worked on the Mk V line, to clean it for him. The reward for this was usually a couple of bob.

I knew that Ralph was off work with a badly sprained ankle, suffered when playing football on the unmade ground outside road test, and it was time for Percy's car to be cleaned. Immediately, his manner changed, and with a smile he reached up, put his hand on my shoulder, and said, "Well, that's good of you. I will have the car parked at the back of the shop. The cleaning materials are in the boot."

In the end, I still hadn't spoken to Vicki, and I spent most of my lunch break cleaning Percy's car, but on the plus side, I had two extra shillings in my pocket.

The next day I was more successful, as Percy was not around, and, to my surprise and excitement, Vicki agreed to accompany me on the trip to London. The weekend of the show arrived and we all had to meet at Coventry station. This meant another early start for me; I had to be suitably dressed and prepared for my first outing to the capital before getting the train from Nuneaton, where I met up with some of my mates. This journey to Coventry was not an unusual thing. Most of the employees at Jaguar (and other manufacturers in the city), came from the outlying areas, such as Nuneaton, Bedworth, Hinckley, and the surrounding villages. One friend of mine travelled from Northampton every day, not much of a journey today, but back then it was a major exercise in transportation.

The train was ready on platform two when we arrived at 8.00am, and Vicki was there waiting. We were quickly engulfed by a crowd of loud and excited passengers. Vicki and I managed to find seats together, and settled down for the journey. Conversation was difficult with the boisterous crowd around, and all we seemed to talk about was our work at Jaguar. I learned a little about her homeland and her family, and I tried to explain mine. We all disembarked at Victoria station and rushed to catch the tube to Earls Court. Everything was so big and so different. It was as though we had been transported to another planet. After our closeted upbringings in rural England, we were not prepared for the rush and bustle of city life.

Eventually, we were deposited at Earls Court station, and following the directions we'd been given, we made the short walk to the exhibition halls. I will never forget my first impressions of that entrance. Programme in hand, I was immediately confronted with bright lights, acres of carpet, and, overhead, all the banners and signs of the various manufacturers. Nothing could have prepared the average visitor for those early exhibitions; for the smell of paint, leather, and oil, the noise of the crowds, or the gleaming displays of the products of the British motor industry.

It was difficult to get near the Jaguar stand, but there, on display for the world to see, was a beautiful pale blue Mk VII, sitting on a crimson carpet, slowly rotating. It was breathtaking. The XK120 in front of it looked just as gorgeous, but there was no doubt that the Mk VII was the star of the show!

Although the Jaguar display was, to me at least, the most striking stand, there were many other attractive exhibits, and the specialist coachbuilders had some marvellous displays too. We spent some hours wandering around the many displays, trying to take it all in. It was early afternoon when hunger drove us to one of the many food stalls, where I had my first hotdog. The rest of the day went by, and, for a reason I cannot remember, we ended up in His Majesty's Theatre, sat in the gods, watching a play called *His Excellency*, starring Eric Portman. I have not a clue what it was about; I even managed to sleep through part of it.

After that, it was a rush to the tube, and back to Victoria where a crowd of very tired but very happy (some too happy from the effects of alcohol) Jaguar employees boarded the train for Coventry.

This was to be my one and only date with Vicki. Shortly after, I heard that she had left Jaguar and returned to Poland with her family. I have often wondered what happened to her.

Winter was upon us, and the five mile trek to my pick up point with Bertie became, at times, a nightmare. I would be deposited in the car park, often wet through and shivering with cold. As quickly as possible I would sneak five minutes for a quick change of clothing (I wisely kept a spare set in my locker). By the time that I arrived home at night, I would have to go through the same routine. How my mother coped with this I never knew. This was before the days of generally available washing machines, and all she had was a coal fired wash boiler, a dolly tub and an antiquated mangle.

At Swallow Road, production of the XK120 was slowly and painfully rising. At least it was until just before Christmas, when industrial relations dropped suddenly in an alarming manner. There were rumours of trouble ahead, most of which seemed to revolve around what was called the 'Coventry rate' of payment.

By this point, the various unions within the works had formed a 'joint committee' to discuss issues, although each individual union still negotiated with management on the subject of pay. There appeared to be more squabbling among this joint committee than ever took place with the management.

I must admit that, to this day, the reasons for the impasse that arose still mystify me. I suspect that the militants among the workforce, seeing the success of the new models and the potential profits that would be deposited in the pockets of the shareholders, wanted their share of the fruits of their labour. Eventually, a mass meeting was called, which was held (with permission from the management) on the waste ground between the factory and Beake Avenue. The senior stewards addressed the assembled, and largely bewildered, throng, and accused the management, and William Lyons in particular, of high-handedness and a refusal to negotiate on the subject of pay rises. A vote was called for the immediate withdrawal of labour, and, following a show of hands, the union deemed the result to be 'unanimous'. I noticed that Mr Shifty Eyes was one of the counters of the votes in favour, which made me immediately suspicious of the union's motives. It seemed that they favoured a strike, come what may; when I, and what seemed like most of the workforce, raised our hands against the motion to strike, no one even bothered to count. "Carried" was the cry. Thus started a depressing six weeks of inaction.

Very few of the people who worked around me wanted, or could afford, to be out of work, but we did not have any other options. Many of us, and I must admit that I was one of them, were prepared to defy the union and continue to work. The threat, however, of being branded a 'scab,' and the potential implication of others refusing to work or cooperate with you afterward, was enough to deter us. I certainly recall that my father's insistence on my union membership and support of the Labour party, led to a great deal of invective between us during the strike. We were told to watch the local papers for information on the progress of management negotiations, which both sides were now forced to resume. It was a very unhappy period, and I remember reading, with great relief, the advertisement in the *Coventry Evening Telegraph* calling the Jaguar workforce back.

There were no winners in this dispute. The only saving grace for management was that they gained some valuable time for getting ready for the inevitable restart. For the workforce, it was empty pockets and a miserable Christmas.

Whilst I have been somewhat critical of union officers, there were exceptions to the rule. Management had learned that a successful shop steward had the potential to cross the divide and 'sleep with the enemy.' The criteria for both was

The production line circa 1951. Mk Vs on the left, XK120s on the right.
(Courtesy Jaguar Daimler Heritage Trust)

NUB 120 was a works XK120 lent to Ian Appleyard to take part in the Alpine Rally. Ian found glory for himself and his wife, William Lyons' daughter, Pat, winning the Rally outright.
(Courtesy John Starkey/Jaguar Daimler Heritage Trust)

similar: a good communicator, ambitious, dedicated to a chosen path, and willing to work within an organised system. Mr Shifty Eyes definitely did not fall into any of these categories. One who did, however, I came to know very early on.

Walter 'Wally' Turner was an operator on the Mk V line and later on the Mk VII. I recall at one time him fitting the boot lid casing and inner boot trim panels on the Mk VII, but he was also our NUVB shop steward. He was a quiet, unassuming man, not in any way militant, but very forthright. He just wanted the best for himself and the people around him. I had no idea at the time that he would cross the divide and become a production plant manager, a post also held by Peter Craig during his meteoric rise to plant director. Another person worth mentioning was Harry Adey. Harry was the NUVB convener for the Jaguar plants. He would also eventually rise to a senior position in personnel on the other side of the fence.

All this marked the start of a particularly bad period of struggle, when labour relations in the motor industry in general were at their worst, a problem that would continue for many years to come. Management itself was not without blame, and there is no doubt that situations were manipulated to suit the circumstances. That said, at last we were back at work. We licked our wounds and carried on the job of building motor cars.

By February, the Mk VII was beginning to trickle on to the assembly lines, and for the first time in my memory, one had sight of three distinctly different models on the same assembly line: the Mk V, the XK120, and the Mk VII. It wouldn't be until the introduction of the Daimler 250 and the V12 that Jaguar would have a mixed-model engine track too.

I was given the job of following each Mk VII through the pre-mount lines, and learning the different requirements that were needed to build it. The Mk V was much easier to build. To start with, the body came to us more or less as a tub, no wings, bonnet, or engine bay to worry about, so access was superb. The Mk VII, however, was just the opposite, with its one-piece construction. To work in the engine bay one not only had to climb over the massive wings, but also negotiate under the propped open bonnet, which was not an easy task.

A great deal of paint damage occurred on the early cars, resulting in the need for protective coverings to go over the wings, which were the most vulnerable. These protective covers were made in the trim shop, and worked well until they had been scuffed around, as inevitably happened. After that they would do more harm than good, as particles of metal and dust got trapped in them and would lead to abrasion marks in the paint which were very difficult to remove. Fortunately, at this time, Jaguar applied enough paint for most of these to be cut out. It would be many years before the problem of paint damage during assembly was almost eliminated.

The bonnet safety latch was another booby trap ready to scar the unwary scalp, and I recall at least one visit to the factory surgery to have a deep laceration treated. After that, I took to wearing a cap because, of course, safety wear was as yet unheard of. Apart from that, the 'big Jag' was not much different to the Mk V, in terms of assembly. In early spring of 1951, we started to assemble the XK120 FHC, to my mind the prettiest Jaguar ever built. I vowed that one day I would own one, but that was still some way off.

For a while now, I had been an avid reader of motoring journals, scrounged when possible, or purchased from my meagre weekly wage of four pounds, two shillings and sixpence. I was particularly keen on motor racing, and followed the early success of the XK in competition with pride. My weak spot, however, continued to be flying, and I still have some of my early copies of *Flight* and *The*

The Jaguar XK120 FHC of 1951. This one was on a camping trip in Dubrovnik in the 1960s. (Courtesy Martin Cutter)

Airplane. I was still a member of the ATC, and had earned my glider pilot wings at Bramcote aerodrome. I discussed all this with Peter Craig, who was sympathetic. He suggested that when the time came for my National Service, I should volunteer for a short commission in the RAF. This would not only provide a more secure income but would also enable me to learn a specific trade which would stand me in good stead when I returned to Jaguar. He assured me that there would always be a job waiting for me, and true to his word, it happened just that way.

My thoughts were soon turning to my 18th birthday, and the arrival of the dreaded military conscription. In early April, being mindful of Peter Craig's advice, I sought reassurance from the company that I would be able to return after my term in the armed forces.

There was no compulsion on the company to promise this if I left on a voluntary basis, but conscription required a commitment to re-employ. With this commitment confirmed, I made my way to the RAF recruiting centre in Broadgate, and signed on the dotted line for four years of service.

I signed on with the desire to be aircrew, but, to my horror, when taking the medical examination, I was declared partially colour blind, which eliminated me automatically. After that diagnosis, my weakness in mathematics didn't matter in the end. Following this gut wrenching knowledge, my next option was ground crew, and I chose to train as an 'electrician air,' which meant that I would be trained to work on aeroplanes rather than road vehicles.

So, in April 1951, I left Jaguar to serve in the RAF. I said my goodbyes to all of the friends that I had made, some of whom I would never see again. I was not to know that when I returned, the company I had worked for since leaving school would not only be located in a new factory, it would be unrecognisable.

As the move happened, people left for new pastures and other people joined. The British motor industry was in a state of flux, and within a few years the city of Coventry, which was once able to boast, with absolute sincerity, that it was the centre of all things connected with the motor trade and its related engineering, would become a city of retail parks. We all saw what was happening to the British motorcycle industry, and blithely said that it could never happen to us. How wrong history has proved us to be!

Chapter 5

1951-1955

The RAF, towing aircraft, and night flying

A new part of my life began at RAF Cardington. I was kitted out in the hangar, where the ill-fated R101 airship was built, not knowing that years later I would return to this historic site. After kitting out, my intake and I were despatched to RAF West Kirby, on the Wirrell, for initial training. There, I learned how to blanco my webbing and gaiters, polish my boots to a mirror finish, march, and handle a Lee Enfield .303 rifle, with which my father had already giving me a head start.

I was rudely awakened to the fact that service life would not be easy, rising at 6am (which I was already accustomed to), followed by physical exercises (which I was not used to), breakfast, kit inspection by the orderly officer, and then drill, drill, and more drill. Marching became a regular routine, and I must have covered most of the camp's surrounding countryside during this period.

Rifle practice on the camp range was my favourite part of service, and I became quite competent with my .303. What all this had to do with the service and maintenance of the RAF's front-line aircraft is still unclear to me.

During this initial training, I remember one of the members of my intake (a conscript), was either mentally unstable or wished to appear so. When told to blanco his kit and get it ready for inspection, he did just that, everything, uniform, greatcoat and anything else that he could find. I thought that the flight sergeant instructor was about to have a fit when he saw the result, the culprit was marched out under escort and we never saw him again.

After six weeks of sheer hell, we were deemed fit to represent the country in the event of hostilities, and were despatched to our respective trade training stations. In my case, this took me to RAF Melksham in Wiltshire, home of electrical, radio, and radar training. Here, I learned not only the basics of electricity, but also how to enjoy the local beverage. I think it was called Black Velvet, a fifty-fifty mix of scrumpy cider and Guinness. It was a delightful drink, until the next morning when your head was invaded by a man with a large hammer, and was only partly relieved by numerous trips to the toilet.

Melksham had a hangar dedicated to aircraft from the Second World War, including an Me 262, the world's first operational jet fighter, and, as its centrepiece, an Avro Lincoln bomber. We used the Lincoln regularly during our training, and I learned how to strip, rebuild, and calibrate its complicated bombsight.

Another three months went by and the training continued. At last, the time came for our examinations. To my surprise, I passed out at the top of my group, and was presented, at the passing out parade, with a set of three bladed propellers to replace the two bladed badge on my uniform sleeve; this promoted me from leading aircraftman to senior aircraftman. After passing out, we were given forms to complete, on which we could indicate our preferences for posting. I chose the Far East as it seemed like a good opportunity to see other parts of the world.

When my posting eventually came through, in the autumn of 1951, instead of Malaya, which it could have been, I found myself on a train to RAF Leeming, near Northallerton in Yorkshire, where I would join No.228 OCU (Operational Conversion Unit).

Gloster Meteor NF11. (Author's collection)

Gloster Meteor NF11 at dispersal. (Author's collection)

RAF Leeming, close to the village of Leeming Bar on the Great North Road (now the A1), has since been bypassed. As far as I am aware, it still operates as a training station. In 1951, there were three squadrons based there, simply called A, B and C, all of which operated the latest Meteor NF 11s, which were all-weather night fighters. The NF 11 was a two-seat development of the F9 single seat fighter, the centre sections of which, in the early days of jet fighter production, were built and repaired in the very factory that I had just left – the 'Manchester shop' at Jaguar's Swallow Road!

Fitted with the latest (at that time) forward seeking radar, long range tanks, and armed with 4 x 20mm cannons, the NF 11 was the RAF's most modern night fighter aircraft.

Where two-seater night fighter aircraft were concerned, the purpose of the OCU was to bring the pilot and navigator together for the first time. The theory was that they would both pass their individual courses before being trained on how to work together. During this training, an aircraft would act as the enemy, while in another aircraft, the pilot and navigator team would practise using the seek radar, and vectoring the pilot to a firing position. After successfully completing this training course, the crews would be posted to the front-line operation squadrons. The conversion course lasted for about ten weeks and part of that would include night flying exercises over the North of England and the North Sea. Night flying support was quite relaxed. All major repairs and maintenance were carried out during the day, so once we had seen the squadron off, there was little to do until they returned, when the aircraft would be checked over and refuelled, ready for the next op.

In February 1952, King George VI died. A full dress mourning parade was held at the station, after which we were told that five of us would be chosen to be part of the RAF honour guard for the funeral at St Paul's cathedral. To my surprise, I was selected as one of the five. We were issued with new uniforms for the occasion, and drilled on procedure for the day. Very early in the morning of this sombre occasion, we were trucked with a police escort to the capital, and taken to the guards' barracks to be dressed and inspected. Two hours before the cortège was due, we lined up at the cathedral steps wearing white (not the traditional RAF blue) webbing, and as the gun carriage carrying the king's coffin approached, we reversed arms and bowed heads as instructed. We seemed to spend hours in this position, but eventually it was a very tired and weary group that made its way back to the barracks to collect our gear and be trucked back to the station.

At last, I was really beginning to enjoy life in the RAF, as the work was never the same. During the daily routine of service and maintenance, there was time for relaxation and sport. Sunday was the only free day, and even then there was the mandatory church parade in the early morning.

I was yet to possess a driving licence, and car ownership was still a few years away, but I had made a few firm friends, one of whom was a young Scot from Perth named Bill McOmish. On his 18th birthday, his father presented him with a bullnose Morris Oxford. He was allowed to keep this on the station, and on Saturday evenings we would charge up the Great North Road to Northallerton, a market town some ten miles away. Since this was before the day of breathalysers and speed limits, we were often in no fit state to drive back, so we would take it in turns to drive. The rest of the party would decide who was the most sober, and he

Squadron A's Gloster Meteors on the flightline at RAF Leeming, 1952. (Author's collection)

would drive, licence or not. Today, this sounds very wild and irresponsible, which it probably was, but the Morris was flat out at 50mph with five up, and there was very little traffic on the roads.

I remember one of Bill's party tricks, which was particularly fun if we had a stranger in the car; he liked to take off the steering wheel (attached to the column with a large wing nut) whilst driving along. He'd hand it back to a poor victim in the back, and ask him to drive for a bit. Meanwhile, Bill would be holding on to the end of the column with his other hand to keep the vehicle more or less in a straight line. The reactions from the back seat had to be seen to be believed!

The commanding officer (CO) of all three squadrons, A, B, and C, was Wing Commander Toppham, who turned out to be the brother of Mirabel Toppham, the owner of Aintree race course. He had a beautiful, aluminium-bodied, AC Buckland saloon, which was his pride and joy. One day, 'Chiefy', our flight sergeant, called me in and told me that I was to report to the CO's office. With some trepidation, I did so, nervously pondering what I had done to necessitate such a call.

At once, Toppham put me at ease, and said that he wanted to discuss a personal matter. He had been going through my file and discovered that I had worked at Jaguar, and asked if I was interested in motor cars. When I replied, "Yes," he said, "Good, how would you like to look after my car in your spare time, you know, clean and polish, check the oils, etc?" Well, I could hardly say no, but I did point out that I did not possess a driving licence yet. He said that it did not matter as I would only be able to drive the vehicle within the confines of the camp anyway, and he offered me a small retainer from his own pocket. This arrangement allowed me to learn to drive properly, as well as earn a little money. I ended up spending a great deal of time behind the wheel of the AC. Very often, I would find myself returning salutes from junior officers, who thought that they were saluting the CO; some were not amused when they realised that it was a mere 'erk' behind the wheel.

At one point, I was given a three-day pass, and decided that I would go home to see my parents. Despite many hours of driving, I was still without a licence, so late in the evening I stood at the side of the Great North Road, thumb in the air, hopeful for a lift. The very first vehicle that came along screeched to a halt, and I found myself staring at a very streamlined and powerful sounding sports car, which immediately reminded me of the XK120. I was staring at a silver BMW 328, the first that I had seen since my trip to Earls Court, where I had admired the same model on the BMW stand. A cultured voice with a strong foreign accent asked where I was going, and when I said Coventry, he told me he was on his way to London via Rugby, and that he would be happy to drop me as near to my destination as possible.

I have vivid memories of that journey; I remember the noise of the exhaust, the sweep of the headlights, interrupted only occasionally by the glare of oncoming vehicles. We chatted for some of the time, and he was very interested when I told him that I had worked at Jaguar before joining the RAF. It turned out that he was employed by BMW and was visiting England to talk to potential dealers. He was kind enough to deviate from his route and ended up dropping me off in the centre of Coventry, at Pool Meadow, just in time for me to catch the last Midland Red to Nuneaton and home. I never did find out who chauffeured me that night.

Even without a civilian licence, I persuaded Chiefy to put me in for a test as an internal driver, which, when issued, would enable me to drive the squadron's three-ton Bedford truck and the David Brown tractor, which was used to move the aircraft to and from the hangar. It was not much of a test and only consisted of some general mechanical questions, and a drive through the gears around the MT (motor transport) yard, followed by reversing and parking.

The hours of practice with the Wingco's AC came in handy, and (when not engaged in my proper job as an electrical engineer air, dedicated to working on aircraft rather than ground vehicles) I became the squadron's reserve transport driver.

My favourite job was with the David Brown tractor, towing the Meteors out of the hangar and on to the dispersal pad prior to operations, and then back in again after flying was complete. The towing bar was similar to those I still see in use at airports today. It consisted of two tubes one inside the other, with a large compression spring between them, which took up the initial starting load and acted as a shock absorber when travelling. One end of the bar attached to the tow hook on the tractor, and the other was forked to fit around the nose wheel of the aircraft, where it was attached to the wheel using two spring loaded pins. The whole assembly was designed to 'break' if the load became too great, in order to avoid damage to the aircraft's undercarriage.

A broad yellow line was painted down the centre of the hangar to mark where the driver of the tractor should keep the nose wheel to ensure the wings would clear the hangar doors. The doors were opened by hand using a very large crank handle. Here, two white lines were painted to show how far the doors needed to be opened, again, to ensure the wings would clear them. Across the hangar entrance was a shallow drainage trench covered by reinforced steel drain covers. There was quite a bump at this point, which always taxed the load sensor on the tow bar.

One particular day I was towing out an NF 11, which had just completed a minor service, including a retraction test of the undercarriage. The tractor passed over the bump at the hangar entrance, but as the undercarriage of the aircraft hit the bump, the main wheels, momentarily released of all weight, suddenly retracted, and the whole thing fell to the floor.

With several tons suddenly imposed, the tow bar broke, as it was designed to do, and the tractor, with me desperately clinging to it and now relieved of several tons, shot off like a startled rabbit, straight for another parked aircraft. Very fortunately, I was able to bring the tractor to a halt before any further damage could be caused.

Search and rescue Mosquito Mk IV of squadron A, RAF Leeming, 1952. (Author's collection)

There were a few seconds of deathly silence before all hell let loose. The Meteor was stranded like a beached whale across the hangar entrance. The auxiliary ventral tank had been punctured as it collapsed, and AVTAG aviation fuel was pouring into the entrance drain. Chiefy had now appeared, so angry he looked as if he was doing some kind of a war dance and having a fit at the same time. The man in the pilot's seat, who had control of the brakes, was sitting with his mouth open. The station fire brigade was quickly on the scene and pumped tons of foam into the drain to neutralise the fuel and reduce the risk of fire. The outboard drop wing tanks had both been crushed, but fortunately were not leaking. The main problem was that the aircraft was now blocking the hangar entrance, which restricted access to one door, a problem that would have to be solved quickly.

The station's seven-ton crane was called in, and with slings around the fuselage, the Meteor was lifted, the damaged tanks removed, guided back into the hangar, and placed on jacks.

A major incident such as this called for a board of enquiry, and everyone involved was questioned. The brakeman and I were cleared of any responsibility, but the poor fitter, who had carried out the retraction test and failed to notice that the undercarriage had not locked down, was for the high jump. He got a severe reprimand, a reduction in rank, and a posting out, and considered himself lucky! The Meteor was out of action for at least two weeks, as a full airframe check was required to assess for any structural damage.

Something similar happened some weeks later. We had, in the corner of the hangar, a lone De Haviland Mosquito Mk V, which was generally used for search and rescue if we lost a Meteor. On this particular day, we were told that one of our aircraft had ditched in the North Sea. The crew had been seen to eject safely, but would need to be rescued quickly. The 'Mossie' would coordinate the search, and direct a rescue ship to the stranded pilot and navigator.

It should be noted that the wing span of a Mosquito is some four feet greater than that of an NF 11. For this reason, a further set of lines were painted on the entrance floor to enable the door opener to make an allowance for this.

Once again, I am haring down the hangar, with the Mossie on tow from its tail wheel, when suddenly there is an almighty shout as Chiefy realises that the doors are not open wide enough. But it was too late; we now had the only clipped wing Mossie in the RAF, as the doors neatly took off two feet from each side.

Again, panic ensued, until we heard that the Mosquito was no longer required, as a Coastal Command Shackleton had been scrambled from RAF Kinloss and was already on station. They had found the crew (none the worse for their North Sea bath), and had dropped a life raft to support them until the Navy arrived.

Meanwhile, our Mosquito was towed back into the hangar to await repairs. The resulting enquiry found the man on the doors guilty of gross negligence. Aside from these two incidents, my time spent driving the tractor was mostly smooth sailing, and I enjoyed it very much.

Our squadron leader made a habit of flying himself, mainly to maintain his flying hours, but also to carry out checks on the crew training programme. Occasionally, one of us would be lucky enough to be offered the passenger seat in the Meteor's tandem cockpit. When I was asked to accompany him during a night flying session, I obviously jumped at the chance; who wouldn't? So, wearing the required flying gear, a suit and mae west, I was strapped to my parachute/dinghy, and then to the navigator's seat of an NF 11.

I had already become competent with the operation of the ejector seat during my training, so the CO only had to explain the procedure should we have to ditch over water, which gave me food for thought.

Since then, I have flown many times as a passenger in modern airliners, and I have experienced the effect of acceleration in a powerful sports car, but neither compare to the sudden kick in the back that you get when the brakes are released on a jet fighter at full throttle.

We were soon airborne, flying due south in pitch-darkness and thick cloud, as we climbed away from Leeming. When we broke through the cloud base, my world became bathed in moonlight; a wonderful sight!

The CO gave me a running commentary on our progress as we swung south-east over Lincolnshire, and eventually over the Wash and the North Sea. The cloud had now cleared, so we dropped down to 4000ft, and turned north to parallel the English coast. As we banked, I was able to look down and clearly see the sea below. I immediately wondered how cold it would feel if we had to ditch in it.

We were travelling at 500mph, but it was almost as though we were standing still. As we continued up the coast, the CO pointed out the lights of Edinburgh and, after a short while, Aberdeen. Just after this, we flew over the mountains to the west coast of Scotland, before turning south and dropping down to RAF Kinloss, where we practised circuits and bumps before landing.

The CO said that we would be off again in about an hour, and disappeared to the officer's mess. I was taken to the sergeant's mess (by now, I had acquired the rank of sergeant technician) for a welcome coffee and supper. Before long we were off again, heading south once more. The CO explained that he would be dropping down to sea level to act as a target for the mainland search radar station. I do not recall the exact height we were flying at, but I felt as though I could have reached out and touched the water. The sensation of speed was incredible as we flew past Blackpool at over 400mph. I remember seeing the lights along the seafront to my left, and looking up at the tower. It all reminded me of *Gulliver's Travels*, with little houses for little people. All too soon, we were turning east again, to begin our descent into Leeming's airspace, where we landed at about 2am. It was a totally unforgettable experience, and I have carried it with me ever since.

Life was never dull in the RAF, but I must admit that by the autumn of 1954, my thoughts were now turning back to civilian life. I was well aware (through the reading of motoring magazines) of Jaguar's move to Browns Lane, and the successes of the C-Type and D-Type at Le Mans. I still did not possess a car, but decided that it was time I got a licence in anticipation of that event. A friend of mine owned a 1936 Wolseley 10 saloon, in quite good nick, and I persuaded him that it would be a good idea for me to use it as a training vehicle. In return, I

Another view of a search and rescue Mosquito Mk IV, RAF Leeming, 1952. (Author's collection)

would fill the fuel tank. By now, I thought myself to be a competent driver, even though I had rarely driven on public roads, and only when accompanied, as per the requirements of my provisional licence. My pal agreed, so after about three weeks of practice, I applied for the test and was given a date, time, and place.

The town of Northallerton was the chosen venue, and shortly before the appointed time, I presented myself at the council offices, where the test centre was situated. To my dismay, it was market day in the town, which trebled the amount of traffic, and congested the roads around the town. I was also concerned, when introduced to my test inspector, to see a rather large, pompous-looking lady carrying a Pekinese, and wearing a straw hat decked out in flowers and fruit. I decided that this was not good and, combined with the traffic problems, I was now twice as nervous. She got in the car, and sat with the miserable little dog on her lap.

Nevertheless, the first part of the test seemed to go quite well. I had carefully completed the stop/start, reverse, and three-point turn manoeuvres. I thought that I had done a good job, but the reaction from the plump lady was totally neutral. Then came the emergency stop. She explained the procedure, and I was returning back through the market square when a portly gentleman stepped out into the road right in front of me, followed by a sheep. As you can imagine, I stood on everything.

The brakes on the Wolseley, despite its antiquity, were good, and we screeched to a halt. Unprepared for this, the plump lady was thrown forward (this was before seatbelts, remember), and banged her head on the windscreen, dislodging her hat, as well as the Pekinese, which then proceeded to hurtle around the inside of the car like a Tasmanian devil, yapping its head off.

Meanwhile, the portly gentleman and his sheep had continued to cross the road as though nothing had happened, and had disappeared. When calm had returned, I was told to drive back to the test office. The tester, having recovered from her ordeal, led me to the office, and gave me a real dressing down, quite unfairly I thought, before issuing a fail notice. She told me that I drove too quickly and without due care for other road users. It took me some time to get over this experience, and it would be another year before I finally passed the driving test.

At last the time was approaching for me to retire from defending the nation and return to civilian life, although I would have to stay on the active reserve list for some time. There was, however, one final action for me to take before leaving the RAF.

I was asked by the CO if I would like to participate in a new incentive for those about to be demobbed. This consisted of two options: the first was to be a guinea pig at the chemical research station at Portland Down, and the second was a parachute training course at RAF Cardington. The first option was a non-starter; I had no intention of taking any sort of action which sounded that dangerous, which, of course, is quite laughable when you consider that the second option required one to jump from a fast moving aeroplane, and hopefully sail serenely to earth suspended from a nylon sheet, which only minutes before you had been sitting on. Nevertheless, that is what I chose!

I found myself back in the very hangar that my service life had started from. The old airship hangar at Cardington was enormous, and provided an ideal location for pre-jump training. This involved leaping from a high platform while attached to the end of an elasticated rope, which gave enough retardation to enable the jumper to practise falling properly. The next step was more serious. This involved jumping from a basket suspended from a barrage balloon at about 500ft. I will always remember the instructor's voice when we reached that height and stabilised; "Well chaps, there is only one way down and it is through that hole in the floor. Anyone left here with me when we descend will have failed." There was no time to argue or hesitate, and within seconds I was floating over the countryside of Bedfordshire; a marvellous experience!

Somehow, the final testing was simpler. Despite the height and speed of a DC3, jumping was much easier, and I really began to enjoy it. The end of the course was anticlimactic. The passing out parade was as formal as ever, and I was presented with my parachute wings by an air vice-marshal. The only downside was that I would not have a uniform to attach them to, because within 48 hours

I had handed in my uniform, drawn my demob suit, and was on my way back to Coventry, civilian life, and Jaguar!

Chapter 6

1955

Jaguar at Browns Lane, 24 Hours of Le Mans, and my first car

When I returned to Jaguar in May 1955, it was to a new factory, with new faces, and new models. The Mk VII M was now in production, along with the XK140 Roadster fixed head coupé, and about to be introduced was the drophead coupé. The much larger premises, with everything now under one roof, had led to the expansion of all departments, and the introduction of new ones as more processes were taken on in-house.

New to me was the XK body shop, which was beside the mounting track. Here, the raw, single skin outer panels of the doors were married to their inner skins. An assembly jig held the inner bulkhead, front wings, and outer bulkhead together prior to welding. The outer skin of the boot lid was also attached to its timber frame. When complete, the whole body assembly was sent down the short lead loading and finishing track before entering the paint shop.

The Mk VII body, produced in Oxford by Pressed Steel, was arriving daily by road transporter, six to a load, in bare metal (referred to as 'body in white'), which required degreasing by hand before it could go through the paint process. Later, bodies would be delivered in primer to avoid this messy and time-consuming work. A buffer stock of Mk VII bodies were kept on site, and could be found in the most surprising places, as outside storage was still at a premium. The assembly lines were very much as before, although longer.

The bodies were conveyed using tracks that had been transferred from Swallow

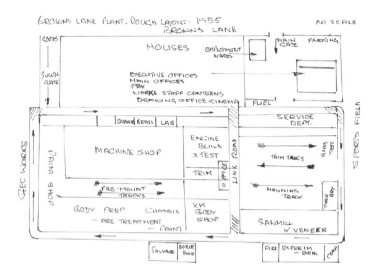

A schematic view of the Browns Lane factory, circa 1955. (Author's drawing)

Road. Having passed through the paint shop, the bodies were mounted to a series of four-wheel steel trolleys. The lead trolley was hooked on to a chain, driven by an electric motor, which pulled the trolley along. The following trolleys were connected by a tubular draw bar, and as the first body reached the end of the pre-mount line, that trolley would be removed, making the trolley behind take over as the puller. The empty trolley was then taken back to the paint shop by hand, ready for another body to be mounted.

It was still a mixed assembly line, with XKs and Mk VIIs running together. Only when they reached the trim tracks would they be separated. Peter Craig was now a senior foreman, along with George Lee (trim track), and Alan Smart (mounting). Also always around was Bernard 'Bert' Hartshorn. At the time, he was assembly line manager, and was another who had followed the company from Blackpool. I would get to know Bert quite well after his retirement. When we wound up living close to one another, we would meet at the local and talk about old times.

Browns Lane was, for some reason, built as two factories of roughly equivalent size, which were separated by a road about 20ft wide. At some point, probably during the war, the two had been linked together by a roof with closing doorways at each end, I suspect to create more space. This area, forever known as the 'link road,' was one of the main thoroughfares in the plant and off it, on the first floor, were the production offices.

Later, when Browns Lane was closing down, I took part in a film made by Jaguar Daimler Heritage Trust (JDHT) to celebrate the history of this renowned plant, and as I walked around amid the ghosts of the past, I was able to find the wall that once supported these offices. When the factory was gutted, so that the modern assembly lines could be installed, everything else disappeared.

It was now well into the middle of 1955, and there was much talk of the new model which was due to come into production. The XK/Mk VII pre-mount track was separated from the machine shop by a half glazed steel partition. There was a space of about ten feet between the pre-mount track and this partition, not enough for an assembly line, and yet this was where the new line would be built. One morning we were intrigued to see a gang of contractors starting to move the nearest machines away from the partition. They were closely followed by our maintenance department who then pushed the partition back. This took space away from the machine shop and provided the extra four feet needed to fit in a new track. However, the line would be heading directly towards the offices, so a dog-leg was put in the overhead gantry, which would allow the completed body to be carried around the offices, across the link road, and on to the mounting track.

We were told that the new conveyor would be a modern, overground, continuous, twin-track system, with each body separately mounted, and not reliant on the one in front for its movement. When it had arrived and been installed, it was just as we'd been told. But it was far from new; never let it be said that Bill Lyons spent money unnecessarily.

Whilst most of our competitors would have sourced a completely new plant, Lyons had gone down the road to Torrington Avenue, where the old Mulliner factory (now owned by Standard-Triumph) was being stripped out, and purchased the assembly line which had been used to build, among others, the Triumph Herald convertible. The whole track was stripped out and shipped to Browns Lane. This track did sterling service through the life of the Mk I, Mk II, and S-Type, but was forever breaking down. It was also raised about 12 inches off the floor, making it downright dangerous.

The main engine assembly shop at Browns Lane, circa 1956. (Author's collection)

The track speed could be varied by a manually operated controller, which was preset to the speed required for the daily production figure. Unfortunately, the load on the track was never constant, and with the varying weight of operators working on the bodies, plus the addition of components and supply voltage, the speed would always vary, too. This would drive our shop steward mad, and while he would accuse the management of deliberately doing this to suit them, it also gave him a good excuse to be away from his operation.

The steward, as the union representative, would be the one to agree the track speed, depending on the number of vehicles that management wished to produce in a typical shift, and individual jobs were timed to suit that. Payment was made at an agreed figure per car taken off the track. However, it often suited management to slow down the process or speed it up.

For example, if the agreed production schedule called for 30 cars to be built during a shift, and the track speed was set to accomplish that, a slightly increased speed could provide management with a 'free' car each shift. At the first sign of any change the steward would immediately call for a meeting with management, and would therefore be absent from his operation to deal with union business.

Whilst the new track was being laid, I was intrigued to see a set of large rollers being installed at right angles to the line, laid across the track and immediately preceding it. This was the new pre-mount roller track, raised about two feet off the floor, and long enough to hold about eight bodies, side by side, door to door.

The freshly painted bodyshell would be delivered by trolley to the start of the roller track, and then lifted onto it. Several operations were carried out here; mainly referred to as black or dirty work. The sound deadening felt was attached to the inside of the roof, the doors, the boot, and under the bonnet. The body's electrical wiring harnesses were pulled through the sills using draw wires, which was a difficult operation because of obstructions in the sill. Eventually, the draw wires would be built in during body assembly at Pressed Steel, and then discarded once used. The bodies were then pushed by hand along the roller track, until they could be lifted and dropped on to the moving line.

Another operation, which was a new departure for Jaguar, was the method of fixing the headlining over the door apertures. On previous models, plywood shapes were drive-screwed to the body, which the headlining would then be tacked to. The whole thing would then be covered by the polished wooden capping. On the 2.4-litre, a square rubber section with canvas layers was let into a pre-formed channel in the body, and then fixed in place using hardened spiral nails through the metal channel. I do not think that much thought had gone into the implications of a nail breaking and hurtling around the inside of the roof, which happened on a regular basis.

In the autumn, Peter Craig informed me that I would be moving on to the new 2.4-litre saloon assembly line, as soon as there were enough bodies and components to justify it, although it would be some time before the track would run with any sort of regularity. It was the XK120 all over again. It would be years before Jaguar finally got their act together and opened a pre-production department, which would build cars on a short assembly line, to prove the operations and iron out the snags, before putting the car into quantity build.

It was around this time that I finally bought my first car, although to call it that is a slight exaggeration because it would take me the best part of two years to build it. I became the proud owner of a 1933 (the year that I was born, which must have been some sort of omen) M-Type MG. This little jewel, with its boat-tailed back, and single OHC four-cylinder engine, came to me in four large boxes, with a chassis, and five wheels and tyres. For this, I parted with the princely sum of £25.

Before I could start work on turning this collection of spare parts into a roadworthy vehicle, I had to build a garage at the bottom of the garden. This did not enamour me to my wife, whom I had married in 1951, or my neighbours. Haynes manuals did not exist then, nor was there much information available from the manufacturer, so it was all trial and error. It also didn't help that most of the body was missing, apart from the scuttle, doors, and bonnet. I constructed the whole of the rear end just by looking at photographs and other cars.

June 1955 saw my first trip to 24 Hours of Le Mans. With a couple of like-minded friends, we made the journey to the Sarthe in someone's Ford Thames van. We camped in a small tent, with no facilities, on the outside of the circuit, roughly where the Porsche Curves now exist. To make the 4.00pm start, we made our way to the banking opposite the pits, in the hope of seeing more of the action. From the very beginning, it was clear that the Mercedes 300SLR and the Jaguar D-Type were well matched. While the Mercedes seemed to have the edge in terms of acceleration, Mike Hawthorn's Jaguar appeared to have the advantage with

A 1933 MG M-Type - my first car. (Author's collection)

The works Ferrari F1 cars at Aintree, 1957. (Author's collection)

braking, due to the performance of its disc brakes, and despite the unique air brake on the back of Juan-Manuel Fangio's Mercedes. For lap after lap, they drove close together, swapping the lead as they came past the pits.

The circuit, of course, was much different to what we now see; the Porsche Curves and the Ford Chicane did not exist, and the pits were much closer to the track, with no protection for mechanics working on a car. The Mulsanne Straight and Indianapolis corner remain much the same today, but with run off-areas rather than sand banks. Arnage is probably the corner that is the least changed, and I suspect that it is still quite possible that a car approaching there with no brakes or steering could go straight through the flimsy boarding, over the Arnage Road, and end up in the middle of the village!

That evening, at 6.20pm, we witnessed an accident that has since been dissected many times. I can recall the sight of the D-Type, slowing to enter the pits, a flash of silver, the awful rending of metal, and fire on the terraces around us. A terrible groan came from the onlookers, and it was obvious that this was a major incident. A car, or parts of it, had cleared the protective banking, and gone into the crowd. We knew it had to be Pierre Levegh, as he had been close behind Hawthorn for some time. There was such confusion that the race became secondary, as the French police, officials, and first-aiders tried to cope with the unprecedented circumstances; it was only much later that we became aware of the enormity of the accident and the subsequent loss of life. We were keen to leave the terrible scene as soon as possible, along with many others, and made our way along the circuit, first to Mulsanne, and then to Arnage. Of course with only a French commentary (no radio Le Mans then) and no television screens, the aftermath of the accident and the progress of the race were guesswork. Late in the evening, it seemed that the Mercedes team had gone missing, but after a few laps we realised that they had been withdrawn from the race. From that point, we lost interest; the race became boring and we left early to start our return trip to Calais. We never got to see a subdued Mike Hawthorn take the chequered flag.

When we returned to work, the tragedy was made even more heartbreaking, when we heard that John Lyons, William Lyons' only son, had died on his way to the Le Mans event. This was followed by many rumours floating around the factory. Most people thought that William Lyons would give up, and with no heir to take over from him, sell the business in his grief. This lead to much speculation as we wondered who the new owners might be. But, as we all know, Bill Lyons was made of much sterner stuff, and threw himself into the business as though nothing had happened. Heaven knows how he and his wife coped with the loss of their only son.

Another memorable motor racing trip that year was in July for the British Grand Prix at Aintree. Things did not go so well for Jaguar in the supporting sports car race, when the lone works D-Type driven by Mike Hawthorn came fifth (behind the entire Aston Martin team of DB 3s), and Ninian Sanderson was sixth in the Ecurie Ecosse D-Type. I did, however, get to see my favourite driver, Stirling Moss, finally beat the master, Fangio, both in identical Mercedes.

Back at Browns Lane, things were gearing up for the 2.4-litre production as bodies from the paint shop became more plentiful. Number two pre-mount was now running on a start/stop basis, and it was odd to see the line with huge gaps between the bodies. When a completed body reached the end of the pre-mount track, it would have to stop and wait until the trolley carrying the engine and suspension components could be phased in on the mounting track.

While building the 2.4-litre saloon, Jaguar experienced several firsts, the least of which was the windscreens. Rather than the multiple piece windscreens we were used to, this saloon had a single piece windscreen, which required new assembly methods.

At times, Pressed Steel's build quality was abysmal, so the screen surround would have to be battered into shape with a large lump hammer, and have its edges filed straight. The screen and its rubber moulding were assembled on a bench trackside, using Sealstic sealant. On the outer edge of the rubber moulding was a groove that the car body would eventually slot into. Into this groove, a drawcord was inserted.

Fitting the windscreen to the car was a three-man operation. A special wooden bed had been built, which dropped over the transmission tunnel. The first operator would lie on his back on the wooden bed while the other two would assist through the doorways. With the windscreen roughly in place, the first operator would now place both feet on a felt pad against the glass, and gently push. Meanwhile the other two, at the right moment, would pull on the draw cord. With a bit of luck and just the right pressure, the screen would pop into place. Too much pressure, and the glass could break. Not enough, and the flange of the rubber moulding would miss the body, which meant refitting the cord and starting all over again. It took some time for the operators to become competent at this, and initially glass breakage was very high, as was damage to the paintwork around the screen aperture.

The next trial was to fit the chrome windscreen finishers. These were made in-house, in the brass shop, and chromed in the plating shop. They were never quite the right shape and they required lots of filing and twisting to get a reasonable fit. Even then they had to be stuck in place with Bostik adhesive, and taped over until the adhesive had set. No two were alike, and would not be interchangeable car to car. The same team of operators were responsible for the fitting of the windscreens as well as the finishers, right through to the end of the Mk II production, and became very skilled at the job.

Another operation that had a tendency to cause damage, was the fitting of the battery on right-hand drive cars. On these, the battery had to be lifted over the offside wing, guided at an angle between the open bonnet and the brake and clutch master cylinders, and down onto its moulded tray. Both the tray and the battery would often be damaged in the process, after trying to get the heavy battery into the small space. Yet another first for the 2.4-litre line, was the designation of a 'finished' area at the end of the track. About four cars back from the lifting point, there was a cut-off point. This provided a buffer zone of completed cars. The theory was that by this point all operations and inspections should be finished. If an operation drifted past this cut-off point and into the buffer zone, additional labour would be put on to pull the operation back.

The 2.4-litre Jaguar Mk I. (Courtesy John Starkey/ Jaguar Daimler Heritage Trust)

That was the theory. In practice, though, it rarely worked. For one thing, it was not productive to have labour standing around, waiting for a situation to develop. The few spare operators we had would be covering absenteeism and other unforeseen circumstances, like the dreaded shop steward not being around.

Most operators would work their way back up the track, by doing their operations quickly, to create some spare time for themselves, but if this happened too often management would want to know how it was possible and look at having that particular operation retimed.

Me at Aintree with a D-Type Jaguar. (Author's collection)

Operation timing was always a troublesome thing. At that moment in time, everyone was paid on a piecework rate, agreed by the union representative and junior management, based on the number of cars to be produced in a shift. The track speed was then set to reflect this, with some leeway in case something were to go wrong.

As a new model or operation came on to the tracks, it would be timed by the shop steward and the work study, time, and motion department. They would then jointly agree (and sometimes disagree) on the time to be set for the job. The times for all operations would then be added together to give a total build time for the car.

The conflicts occurred because it was in the management's interest to have the time set at or above the normal judged working rate, in order to keep production numbers as high as possible. On the other hand, it was in the operators' and shop stewards' interest to slow the job down as much as possible, knowing that, when working at the proper rate, spare time would be created.

The problems this system created were never properly solved until the end of piecework. If management suspected that a set time was wrong, they would insist on a retime with a different operator, which the union representative would resist. I was often called in to provide the comparison timing, and many times crossed swords with the steward, when I would be accused of being a management 'lacky.'

As a rectifier, apart from reworking an operation, I was also expected to fill in for operators who were absent, which was usually the steward. As he had the easiest operation on the line, I would quickly work the job back to the stage where I would be waiting for the next body to come off the conveyor. I took great pleasure in doing this with Mr Shifty Eyes operation, which would drive him berserk as it made him look the fool that he was. It culminated one day in the foreman insisting that the operation was retimed. Mr Shifty Eyes could hardly refuse to be timed, because if he did, they would time me instead, and he knew that I was quicker. After that it was quite satisfying to see him having to work for a change, though my relationship with him went downhill even faster.

One of the fascinating areas of production at this time was the D-Type assembly line. It was common knowledge that to comply with the Le Mans regulations as a production sports car, the manufacturer had to supply physical proof of the intention and capacity to produce 100 identical models. Most enthusiasts will, by now, be aware of some of the lengths that manufacturers, such as Ferrari, would go to in order to appear as though complying. Needless to say, at Jaguar, this was not the case.

A real production line was set up between the mounting track and the Mk VII trim line. However, it was never considered a true line because it was static. The cars were pushed along by hand, guided by two timber rails set on the floor of the factory. This was enough, though, to supply the proof of the capability to build the numbers required, although, as it turned out, the magical figure of 100 was

The 1953 short-nosed D-Type Jaguar. (Courtesy John Starkey/ Jaguar Daimler Heritage Trust)

never reached. There is no doubt that the original concept would have enabled Jaguar to sell the whole of the D-Type production, but that is not what happened.

As they were completed, the finished, unsold cars were painted in grey primer coat and stored in the tin sheds between the sports field and the main factory, where they remained safe from the disastrous fire of 1957.

The D-Type production was unusual in that, although it was situated in the main factory, it did not come under production control. Instead, the build was kept firmly in the hands of experimental and competition departments.

I used every excuse possible to spend time on the D-Type line, and became familiar with some of the well-known figures who were working there, from both the experimental and the competition departments. I became particularly friendly with Bob Penny, Joe Sutton, and Bill Jones, whose brother, Tom, was working in chassis design. I did not know then that within a couple of years, I would be working in the same departments as them. Later in my life, I would get to know Tom very well and we would meet up with the gang at Le Mans for the 24-hour race. His daughter, Leslie, and her husband, Ray, are now close friends of ours, and once again, we will meet up with them for the Le Mans Classic.

At home, work on the MG was slowly progressing, and in anticipation of the day when it would be on the road, I attempted the now dreaded driving test. This time I passed and became the proud owner of a full licence, still without a car to drive.

I had been gradually collecting parts, mainly from the Jaguar scrap and reclamation department, where any reject or scrap items could be bought for a

Peter Collins' Ferrari 250GT cabriolet outside William Lyons' office at Jaguar in 1955. It was the first Ferrari ever to be equipped with Dunlop disc brakes. (Author's collection)

pittance. So far, I had found a pair of beautiful C-Type bucket seats, trimmed in grey cloth, for 50 shillings each; enough carpet to trim the whole car, which I had sewn and piped in the trim shop; two P700 headlamps; two spot lamps; a pair of D-Type tail lamps; an XK reversing lamp, and sundry instruments and fittings. As you can gather, there was going to be no attempt to conform to MG originality.

While the scrap and reclamation department helped me build my very first car, the system was abused, to the extent that eventually it was closed down. The favourite scam was to buy a scrap item legitimately, with a receipt, then pick up a new one in the factory and walk out with it. Batteries were a common target. Then there were the pilferers; I can recall stories of a man who was found to have an XK crankshaft under his raincoat, and another stiff-legged man who had a camshaft strapped to each leg. I suppose that such things are common even today.

So with a half-built car, 1955 came to a close, and things were pretty good at Jaguar. Production was gearing up on the 2.4-litre; I had now been transferred full-time to the new pre-mount track, and I had joined the Jaguar Aston Martin Club as an associate member, a willing navigator for any driver who needed one in their events. I had more money in my pocket, and was able to indulge my passions of motor racing and air displays.

Chapter 7

1956

A royal visit, Jaguar retires from racing, and preparing for Earls Court

The upcoming year, 1956, would be a significant time for Jaguar, as it slowly became a major part of the British Motor Industry. The New Year's Honours list announced that Bill Lyons had been nominated Knight Bachelor for his services to the industry, for Jaguar's contribution to export in particular; from then on he would be known as Sir William Lyons.

In March, the factory was visited by Queen Elizabeth II and Prince Phillip. As can be imagined, there was great excitement, tempered only by one or two extremists (yes, we had them in 1956 as well), who threatened to use the visit to advertise their particular brand of politics. They were given, quite simply, the address of the Coventry labour exchange; employment was not that easy to find in the mid-fifties and Jaguar was, by now, paying some of the highest rates in the city.

When the great day came, we were all on our best behaviour as the Royal couple toured the assembly lines accompanied by Sir William and Lady Lyons, the Lord Lieutenant of Warwickshire, and the rest of the entourage. I was surprised when Prince Phillip, while on the tour, poked his head under the bonnet of the 2.4-litre that I was working on and asked what I was doing. I told him that I was fitting a heater. In response to my stammered reply, he said, "You are all doing a grand job for England." It was as good as a pay rise!

The same year, petrol rationing was reintroduced following the Suez crisis,

Sir William Lyons, recently knighted, on the left, Queen Elizabeth II and Prince Philip in the centre, and the Lord Mayor of Coventry on the right. This photo was taken during the Queen's official visit to the factory in 1956. (Courtesy Jaguar Daimler Heritage Trust)

which, for a time, played havoc with the second-hand car market. For the first time since the end of the Second World War, petrol coupons were issued, and pleasure motoring was strictly limited. Not that this worried me too much as I still did not have a working motor car, and my bike was still doing sterling service when the weather permitted.

After the terrible accident at Le Mans the year before, the 24-hour race was delayed until August. Track modifications had to be made, and they were not completed in time for the normal June date. The works cars were eliminated early on, as Fairman and Frere had a collision at the esses, which put both D-Types out, while Hawthorn was plagued with fuel injection problems on the remaining car. Although this was eventually rectified, the time lost would see him finish in sixth place. All was not lost, however, as Flockhart and Sanderson drove brilliantly in the Ecurie Ecosse D-Type to win the race. This was to be Jaguar's last official race at the Sarthe for a great many years; in October, the company announced its retirement from racing.

When the time came for the Earls Court show, preparations were made to

A typical wooden dashboard being French polished, prior to being fitted into a Mk I 2.4-litre saloon. (Courtesy Jaguar Daimler Heritage Trust)

complete the cars that were set to be displayed. These were sent down the line as normal, although great care was taken to protect the paintwork from damage during build. Protective covers were made in the trim shop, to mask the vulnerable wings, doors, bonnets, and boot edges.

The cars were then completed off-track in the rectification bay. Most of this work was carried out after normal hours, as labour could not be spared during the usual work day.

During this time, my job was to build and fit the instrument panels using polished woodwork supplied from the sawmill. After that, the exterior chrome work, which was supplied as a complete kit for each car, was fitted. The quality of inspection was unusually high, at last, and there were always high-profile people, such as Lofty England, around, usually wanting to interfere with whatever work was being carried out.

Lofty worked in the service department which was very close to the rectification bay. He would suddenly appear and stand with his hands behind his back, usually wearing a severe frown, as though unhappy with what was happening. I had not forgotten my previous confrontation with him and was still wary when he was around.

One evening, I was in the process of fitting the rear chrome trim on a 2.4-litre, when he approached me and, in his usual terse manner, asked me why I was twisting the trim, clearly thinking that I was going to damage it. I tried to point out to him that the shape the trim came in would not fit the shape of the body; it was necessary to twist and work the trim to follow the contour of the body, otherwise it would be in tension when fitted to the spring clips, making it likely to fly off at some point. He continued to grumble until I handed him the trim and my fitting tool, and asked him to show me how it should be done. He stood for a few seconds, trim in hand glaring at me, then handed it back with the grudging remark, "Well, I suppose you should know what you are doing," and strode off. Strangely enough, I never saw him in the rectification bay again!

Much has been written about Lofty, who in later years I did get to know quite well, particularly after his retirement, but back then I found him to be a thoroughly disagreeable person.

Sir William, on the other hand, although mostly distant from us mere mortals, was always well mannered and willing to listen to your point of view, even if eventually he disagreed with you, something that happened when the two show cars were finished.

That day, we had worked later than usual as the cars were due to be despatched that evening. We carried out the final inspections, and the cars were pushed up the ramps of the transporter that was ready and waiting.

The usual collection of high officials had come and gone, having had their say, leaving Peter Craig and George Lee in charge of the operation. We were all standing back and admiring our handiwork when Sir William and Lady Lyons walked in. They observed the two cars from a distance, called George Lee over, spoke in a huddle, then walked away.

A very angry, red-faced George approached us, and said "Get the bloody green car off the truck. Lady Lyons does not like the colour." Bewildered, I asked what we were expected to do about it, to which George replied "Well, we are going to have to change it, aren't we?" We spent the next hour removing all of the exterior chrome that we had carefully fitted, and watched as the vehicle was wheeled away.

When I arrived at 7.30am the following morning, a pristine pearl grey 2.4 sat waiting for its chrome to be refitted. The rectifiers from the paint shop had spent most of the night repainting the car, the transporter was still there waiting, and we had very little time to refit all the chrome work.

Also watching and waiting was Sir William. After reloading the transporter and seeing it off, he called George Lee over and had words with him before turning and walking away. George told us that Sir William had apologised for the extra work that had been created, and that George was to thank us all for our efforts. Although he was obviously not prepared to thank us personally, he knew what had been involved and made an effort to show his gratitude; that's something Lofty would not have done!

Around August 1956, the production of D-Types came to an end. The space that the temporary production line had occupied was now used to complete the build of the XK140 drophead coupé (DHC). The assembly of the folding hood on these cars was complex, and certainly did not lend itself to being carried out on a moving line. One of the specialist trimmers on this operation was Des Cross, who would later become senior foreman on the Series 1 and Series 2 XJ trim track.

In the space left behind by the D-Type line, several 'special' cars appeared. One was the C/D interim car, which ran at Jabbeke with the XK120, and 'Brontosaurus,' a peculiar-looking vehicle that was Sir William's own idea for a racing sports car. Both of these would eventually be moved to Wappenbury Hall, Sir William's home, before being dismantled. It wasn't long before a more interesting car soon appeared: It was the Mk VII DHC. Although it never saw the light of day, the immense hydraulic-powered hood was trimmed anyway by Des Cross and his mates. As far as I am aware, no photographs exist of this, Jaguar's last attempt to build a DHC on a large saloon, although several independent manufacturers have also tried, most with questionable results. One of the main problems was that taking away the roof and the 'B' post meant that the body lost an enormous amount of strength and integrity. This could only be replaced by strengthening the floorpan structure.

Enthusiasts will be aware that, in much more recent years, the Corsica, a one-off concept car, has been successfully modified by the JDHT, and is now fully

An XK140 OTS, produced by Jaguar from 1954-1957.
(Courtesy John Starkey/Jaguar Daimler Heritage Trust)

functioning. Until this time, it had only been a painted and trimmed rolling shell, with a working hood, and it required substantial stiffening in the floor.

Even when Jaguar later built the two-door XJ coupé, which retained the roof but removed the 'B' post, there would be problems. Owners would find that when jacked from the rear nearside, say, for a wheel change, they might not be able to open the passenger door, because the whole body would be bending in the middle. The XJS convertible, even with its body stiffening, would still suffer from flexing, and required the addition of a stiffening frame under the power unit to reduce this. Manufacturers were quickly discovering that it was not easy to turn a closed unitary shell into an open one.

It was a year of consolidation for the company as production steadily increased. The Mk VIII was now in production, which made little difference to the build programme as the changes were only superficial. The XK140 had taken over from the XK120, and both were selling well, particularly in Jaguar's most important market: North America.

Following the company's withdrawal from racing, the works D-Types were sold to Ecurie Ecosse and raced successfully, particularly at Le Mans, albeit with substantial behind-the-scenes support from the factory. In 1957 there would be significant changes for both Jaguar and myself; some good in my case, and some bad in Jaguar's.

Chapter 8

1957

The XKSS, the fire, and a new position

In January 1957, I at last had four wheels, as the MG was now complete and running (if only up and down the garden). The next challenge was getting it out on the road. I knew when I started this project that I would one day have to solve this problem, but I had been putting it off until the time came.

Fortunately, I had the help of my next-door neighbour who had been following the build with great interest. He was a lovely little Welshman who worked at the Standard factory in Tile Hill.

Our houses were separated by a chain link fence, mounted on concrete posts, with only a four-foot wide walkway down each side. At one point, we had considered gathering the neighbours to help us carry the car over this, but eventually decided against it. Instead, we got to work with pick and shovel, to uplift the posts, and roll the fence back. We were thus able to extract 'Penelope' (as she had become known), before rebuilding the fence. From this point on, she would sit on the front lawn under a waterproof cover, as there was no going back to the garage.

For the first road test, I wore a silk scarf, goggles, and driving gloves (as all proper racing drivers did), but that was not enough to stop it from ending in disaster. I had driven less than half a mile, when a loud bang, followed by a substantial grinding noise and a total loss of power, heralded the failure of the fabric coupling in the propshaft. To say that I was embarrassed is an understatement, particularly as the whole estate was watching. To make matters worse, the only 'power' available was child power, and so Penelope and I were pushed back to the house by a hoard of howling schoolchildren.

The coupling was not that difficult to replace. I used a section of half-inch belting from the local coal mine where my brother happened to be working at the time, and once more I was mobile. On the morning of our first journey to Browns Lane, the sun was shining, but within five miles all that changed and a cloudburst hit us. With no hood, when Penelope and I arrived in the car park, we both resembled drowned rats.

On the way in I met Alf Baker, one of the works firemen, who would eventually become head of security and fire protection. Alf took pity on me, for which I will be eternally grateful, and arranged for Penelope to be parked under the canopy alongside the fire station to dry out. I, in the meantime, had a change of clothes and clean overalls to look forward to.

Late in January, the XKSS was announced. Despite what has been said over the years, this was never intended to be a continuing production model. The company had been, somewhat desperately, trying to dispose of the remaining

The XKSS. (Courtesy John Starkey)

In later years, I built myself a Jaguar XKSS replica. This is the engine I made for it. (Author's collection)

D-Types, which, as previously stated, were now in store. There were attempts to promote them by reducing the price substantially, but there were few takers. It seemed that no one wanted an aging, uncompetitive sports car when there were now cheaper and more modern alternatives, but something had to be done with the unwanted D-Types.

The history of the XKSS has been well charted by others, but the concept, for that point in time, was brilliant. The first step was to take a D-Type and remove the driver's head fairing and the centre stiffener, add a windscreen, a door for the passenger, and fix some side screens to the doors. Next, improve the trim with carpets, add a rudimentary hood, a luggage rack, and windscreen wipers, and suddenly you have a wild, Ferrari-beating sports car for half the price!

There is no doubt that Jaguar would have been able to sell more of these fabulous creations had it been physically possible to manufacture them, but unfortunately, this was not the case. The cost of re-tooling for the bodywork alone, would have made this unprofitable, not to mention the mechanical requirements, the labour, and space issues, at a time when Jaguar was flat out making bread-and-butter road cars. As it happened, fate took hold of the XKSS in a manner which could never have been anticipated.

On the evening of Tuesday, February 12, I was doing what all good rectifiers do: working overtime! One big problem that we experienced on the 2.4-litre saloons was damage to the forward engine bay electrical harness, where it crosses the bulkhead. Its routing was very close to the rear of the cylinder head. This meant that if the body was not perfectly aligned when being lowered on to the mounting track, it would cause contact between the cylinder head and the electrical harness, which would always come off worse.

Lowering the body on to the trolley carrying the front beam, engine, and rear axle, required three operators; one to control the hoist, while two others, one at the front, and one at the rear, would guide the body on to its mounting. The whole process was reliant on the operators watching carefully as the body was lowered. An inch too far either way was all it took to cause contact. Then the only option was to change the harness completely, right through to the instrument panel.

On the night in question, the inevitable had happened, and a car with no electrics, was stranded at the end of the 2.4-litre mounting track. I had changed forward harnesses many times, and had it down to a fine art. If I were able to start just before the tracks closed down, I would be able to get away before 6.00pm. I set off, toolbox in hand, and eventually located the car. It had been pushed off the end of the mounting track, and parked alongside the tyre bay, which, it is important to note, is situated just beyond the end of the mounting tracks.

The tyre manufacturer would deliver the tyres to this bay, while the wheels came from the paint shop, or, in the case of chrome wires, from Dunlop. The tyres were then fitted to the wheels, checked and balanced, before being delivered by truck to the mounting tracks, ready to be fitted to the vehicle.

That night, everything went well with the harness change and the subsequent testing. I was packed up and ready to go by about 5.30pm. I had to walk the whole length of the factory to get back to my work station on the 2.4-litre pre-mount track, which probably took about five minutes. As I neared the track, I heard an unusual noise behind me. It sounded like a minor explosion, followed by a continuous roar. I turned, and to my horror, I could see flames licking across the roof above where I had just been working!

I dropped everything and rushed back to the nightmare scene. The tyre bay was well alight, with thick black smoke and flames everywhere. The insulating panels of the roof, with their north light glazing, had been in place since the factory had been built in the early '40s, and were coated with a bitumen-based paint. As the flames raced across the roof, this melted, flared, and fell as flaming droplets to start individual little fires. It was like a scene from *Dante's Inferno*!

I smashed the glass on the nearest fire alarm, which, as it sounded, only added to the chaos. By this time, most of the production staff had already gone home, but the office staff were still on site, and a crowd soon gathered at the main exit of the lower factory, where the cars left road test. Lofty England, to his credit, was soon on the scene, and he organised people to drive out the cars. Unfortunately, there were problems that we didn't have a quick fix for; not everyone could drive and there was no time for anything more than the most basic instructions, and the cars in road test were not yet fitted with seats, as each tester carried his own temporary seat, moving it from vehicle to vehicle. Young girls were trying to drive cars sitting on the floor, and it was like dodgems at the local fair!

By now, Alf and the works fire brigade had arrived, but were faced with an almighty task. The main roof over the trim tracks was well alight; many smaller fires had started at floor level; tins of highly inflammable solutions and white spirits (used for cleaning) were exploding like small bombs. We could do nothing about the cars on the trim tracks, as they were hooked on to the conveyor and it would have been an impossible task to try to separate them. The paint on the bodies was now burning, along with pieces of trim awaiting assembly. Vaporised petrol from the fuel tanks began to explode, and we had to give up. All that we could do now was stand back and watch as our life's work went up in flames.

I distinctly remember thinking that I was going to be out of a job. The company could never recover from this. By now, the Coventry fire brigade had arrived, quickly followed by units from the surrounding areas, and were pouring thousands of gallons of water on the blaze, which quickly turned to steam, and blurred the entire horrific scene.

Eventually, we were forced to leave it to the professionals, but not before seeing the roof collapsing over the service department. It was a heart-rending sight. There were a lot of heroes and heroines that night, who probably, like me, went home tired, dirty, and shaken, wondering what would happen the next day.

Most workers arrived as normal the following day, a great many did not even know what had happened the night before. When I arrived, I was confronted with a scene of total desolation. The fire brigade had poured so much water over the conflagration, that the lower factory was six inches deep in places. Fire hoses snaked everywhere, and small fires were still being damped down. Apart from the outside walls, the service department had been obliterated, and a number of D-Types in the process of conversion, along with all of the customer cars, had been lost. Weirdly, the clock on the outside wall still showed the correct time.

Most of the trim tracks, the main stores, the tyre bay, and the road test section were a total loss. Fortunately, the fire had been halted, by the heroic effort of the fire services, about 30ft short of the link road. This was just as well, as buried under this road was the main fuel supply tank. Heaven knows what would have happened had that gone up.

It was sad to see the skeletons of what had been almost completed cars, still attached to the tracks, just bare, burnt shells. Most of the employees were sent home and told to watch the press for information. The rest, myself included, were kept on to start clearing up. The burnt-out shells of Mk VIIIs, 2.4 XK140s, and D-Types were fork-lifted out of the factory, and pushed into an untidy heap on the front car park, right outside Sir William's office, ready for scrap collection. One can only imagine what he must have thought whilst looking out of his office window. Not that he had much time for ruminating as he and his fellow directors were seen everywhere, organising the clean up.

It was fortunate that most of the mounting tracks had been saved, and the pre-mount lines were unaffected, as they were on the other side of the link road, well clear of the fire. The roof over the lower end of the factory had completely disappeared, leaving just a tangled mass of burnt steelwork. A meeting of the production managers concluded that using a shortened mounting track and a static trim area would enable limited production to commence.

On Thursday 14 we ran the tracks for the first time since the fire, and, using a start/stop programme, were able to complete a number of cars on that day. Slowly, the situation recovered, though it would never be the same again. For a time we worked, shivering, in the unheated, open-roofed part of the factory. Temporary walls were built to isolate the service area and the remains of the wrecked building. Work commenced on the rebuild of the roof, and the departments affected were hastily relocated to other areas. We had been working flat out when the disaster occurred, and space had been at a premium for a while. The loss of most of the lower factory only added to the problems.

To my knowledge, there has never been a satisfactory conclusion with regards to the origin of the fire. There was talk of a clandestine heater being left on overnight (it was winter, after all), but to my mind, the only thing inflammable in the tyre bay were the tyres themselves, and they are not suspect to spontaneous combustion. Then there was the theory of sabotage by someone with a grudge, which is possible; it certainly occurred at a time when labour relations were at a low ebb.

Production gradually picked up, until eventually we were back to the pre-fire level. The 2.4-litre soon acquired new life with the introduction of the 3.4-litre engine, and, for the first time, optional disc brakes. Jaguar would soon take the motor racing circuits by storm in the recently-introduced category of touring cars. Privately-entered, works-backed (unofficially) 3.4-litre Mk I saloons would

The grim aftermath. A fire damaged D-Type in front; fire damaged Mk VIII saloons behind. (Courtesy Jaguar Daimler Heritage Trust)

dominate the scene until the introduction of the Mk II. Strangely, the term 'Mk I' was never used in the factory, and was only introduced to differentiate between it and the Mk II. At the time, in the works, it was referred to as the 'small saloon.'

A major change in my life occurred in mid-March. I was working away under the bonnet of a 2.4-litre, when a head appeared alongside me and a voice said, "Hello, are you Brian Martin?" The voice belonged to a large gentleman in a grey lounge suit. Now, suits normally meant trouble. Only managers, superintendents, and above wore them in the works, and why would one of those be wanting to talk to an operator on the production line unless there was a problem? My thoughts raced looking for the answer to the next question: what problem?

I replied, "Yes," and the man responded, "I am Burt Tattersall, and Peter Craig has suggested that I speak to you." My God, now Peter is involved, this must be serious. I climbed out from under the bonnet, and shook the proffered hand. I remember the operators around me giving each other some sideways looks. They were probably thinking the same as me: I am in trouble!

It turned out to be a very one-sided conversation, as I was totally unprepared

3.4-litre Mk I saloon bodies on the production line. (Courtesy Jaguar Daimler Heritage Trust)

for what Burt said next. "If you are interested, I would like you to come and work for me in the experimental department. I have cleared it with Peter Craig and he is willing to let you go." Well, to say that I was speechless is a definitely justified. I was being invited to work in one of the most exciting departments in the factory, with some of the engineers that I had already made friends with on the D-Type production line. I didn't even think to ask how much I would be paid, or under what terms I would be employed. The answer came easily: YES.

As it turned out, the salary I would be paid was substantially less than that which I was receiving as a track operator, but at least overtime would be paid, and there would certainly be plenty of that. Later that day, both George Lee and Peter Craig came to see me. George said that I was making the right decision, and Peter, bless him, told me that he had arranged an interview with the engineering department. Three days later, I was called to attend this interview, which was a one-to-one with Burt Tattersall. Despite the fact that he had control of the design, supply, and proving of vehicle electrics, I quickly realised that he knew nothing at all about them, which made the interview slightly bizarre. It was fairly straightforward though. After general questions about my employment at Jaguar, and some asking for my personal information, he asked what I knew about vehicle electrics and wiring.

Over the years, I had familiarised myself with the Lucas wiring colour coding, which, at the time, was commonly used in the motor industry, so not a problem there. When I explained to him that the wiring of most cars was similar to that of an aeroplane, only less complex and half of the voltage, I knew that I had lost him completely.

The next question really threw me: "Could you make a bracket?" "Why would I want to," was my immediate response. "And what has that got to do with vehicle wiring?" He told me that I would have to construct various test rigs as part of the job, and that will require an ability to work with metal. I reassured him that I was very competent at making brackets. Thus the interview ended. I have had several job interviews in my life, but none of them were ever as weird as that one.

That was all forgotten when, two days later, I was told to report to the employment office to collect my new clock card. I then went to the experimental department, taking my tools and personal belongings with me, and reported to Mr Cassidy, the foreman.

On March 29, 1957, I clocked-in on the experimental clock for the first time, and was shown to my new station. The grand title of 'experimental electrical development' described a redundant test cell, about 8ft by 12ft, with two work benches and a couple of lockers. Here, I was introduced to Ron Beaty.

Ron is well-known in Jaguar circles as the brains behind Forward Engineering, and Ron Beaty Cars. When I first met him, Ron was just completing his apprenticeship at Browns Lane, and through a series of events, had been left holding the baby, as it were, in electrical engineering. This was only on a temporary basis, as he was due to take up his post in engine development, hence the need to add someone else in a hurry.

I was to learn very quickly how frustrating my introduction to the department was going to be.

Chapter 9

1957-1958

The experimental department, a Ford van, and rallying

In 1957, the experimental department was located in a small, independent building at the north end of the main factory, though would later become part of the sawmill veneering shop when experimental made its final move to the General Electric Company (GEC) site. In this rather crowded building, all of the various experimental functions took place.

In the main shop there were two fixed ramps, each with hydraulic lifting. Alongside these ramps there was a small machine shop containing a horizontal miller, two lathes, and a vertical drill. Further down the main shop was engine assembly. The body shop, stores, and competition shop, with its own roller door access, were all partitioned from the main shop. The small offices were occupied by Mrs Dixon, Norman Dewis, Walter Rheese, Bill Nicholson, Bill Cassidy, and Jack Gannon, with separate offices for Bob Knight, Bill Wilkinson, and George Buck.

I will try to describe the various functions of both personnel and their responsibilities, and hope that my memory serves me well:

Mrs Dixon was our Girl Friday, looking after everyone and manning the reception desk; Norman Dewis, who needs no introduction from me, ran the experimental road test department, usually with varying apprentices as assistants; Wally Rheese worked with Norman, keeping records and occasionally driving; Bill Nicholson was responsible for general upkeep, and introduction of modifications on the fleet of experimental vehicles; Bill Cassidy, in his role as foreman, was responsible for the machine shop, vehicles, and personnel; Jack Gannon, who was soon to move on to new pastures, was heavily involved with ex-apprentice, Geoff Faulkner, in cooling and air conditioning trials, using an early Mk VII M; Bob Knight was the senior engineer and had overall control of the department; and Bill Wilkinson and George Buck were in charge of experimental engines.

There was also Phil Weaver, who was in charge of the competition shop, with Ted Brooks as foreman; Fred Gardner, my early nemesis, was in charge of both the production sawmill and experimental body design, and worked very closely with William Lyons on the styling of most Jaguar bodies at this time; and Harry

The lads who worked in Jaguar's experimental department at the same time as me. From left to right: George Hodge, service engine wizard; Ron Beaty, experimental engine engineer; Phil Weaver, competition shop superintendent; Jim Eastick, experimental engine engineer; Bernard Viart, author of many books on Jaguar; Joe Sutton (behind), competition shop engineer; Frank Rainbow, experimental engine workshop foreman; and me, experimental electrical engineer. (Author's collection)

Rogers was in charge of the experimental body shop, which turned the designs into reality.

There were some grey areas, where roles would occasionally overlap, particularly between Bill Cassidy and Bill Nicholson. The former was an expert machinist and a good controller of personnel, the latter was a wild Irishman with a temper to match, and drove everywhere at 11 tenths. He was a very competent engineer, and a very good driver. At one time, he had led the BSA works trials team to represent his country in the World Trials, and eventually, after leaving Jaguar, he would campaign probably the quickest road-going MGB ever on the race circuits.

There were a lot of people who questioned Bill Wilkinson's position, since he seemed to know very little about engines, and relied heavily on George Buck. George was a mild-mannered, softly spoken man, with an immense wealth of engine knowledge. Another motorcycle man, George had competed in the Isle of Man TT before the war.

Harry Hawkins, Bill Cassidy's right-hand man, was a larger than life character. Harry had been heavily involved with the early Le Mans works forays, looking after transportation of the cars, and caring for them at the races, as well as after. He was a bull of a man, with a voice and an attitude to match, but it was all show really; he had a heart of gold, and was always willing to help, even if he made it seem a chore. Working with Harry was Frank Lees, one the best welders that I have ever come across. The pair of them could make or repair anything that was put in front of them.

Peter Blackmore, another of the old hands who had been a Le Mans man, was a vehicle engineer along with Stan Hanks. Peter would eventually leave Jaguar to become one of the founders of BWL Motors in Coventry, a company that still exists today, although Peter retired some time ago. At that time, BWL specialised in the service and repair of Jaguars, and did it very well.

Soon to join them in the road test department was Barrie Woods, whose wife-to-be, Pam Joyce, worked in the Jaguar telephone exchange, and was the cause of much heavy breathing among some of the young apprentices. Barrie was once an apprentice himself, but he eventually left Jaguar to join AC Delco and become its European sales manager, with one office in Dunstable and another in Germany.

Another life-long friend of mine, Barrie was always up for a laugh. I remember one particular occasion when Barrie, Jack Brown (a soon-to-be member of the electrical department), and I had been to the Earls Court Motor Show. On the way back, Barrie could not hold on to the mussels he had enjoyed for supper, and decided to leave them right on the hard shoulder of the M1. Then, the headlights of his Cooper Mini failed, and he spent the rest of the journey back to Coventry only two feet from our boot so that we looked like one vehicle!

Frank Rainbow was foreman of engine build and test, and had a wealth of knowledge. He was also a major part of Jaguar's competition programme, and you will find many photographs of him in Jaguar books, working on the works cars at race meetings.

In Eric Dymock's splendid book on the history of Ecurie Ecosse, there is a photograph of Frank working on a suspended 3-litre XK engine which has been removed from the Tojeiro Jaguar prior to the Le Mans race of 1959. This proves that, despite David Murray – the head of Ecurie Ecosse – and co claiming that Bill 'Wilkie' Wilkinson did all of the work on their engines, it was not the case.

Working in Frank Rainbow's department was ex-apprentice, Jim Eastick (a fellow Swallow Road survivor), who ran the test cell next to the experimental electrical department, and Bill Duff, who would eventually team up with Ron Beaty in another test cell.

The competition shop was staffed by people who had been, and would continue to be, responsible for Jaguar's success on circuits around the world. Some I had already met on the D-Type production line, and I would continue to work alongside them on and off for the next ten years. Ted Brooks, Bob Penney, Len Hayden, Gordon Gardner (son of Fred), Joe Sutton, Peter Jones, and of course, Bob Blake, the quiet American.

Many good things have been said about Bob's ability to work with metal, and every bit of it is true. Without his input I doubt whether the E2A or XJ13 could have been built, and it was Bob who taught me how to weld aluminium using one of the early Argon arc welders.

All competition shop personnel wore white overalls, with the Jaguar logo on the back. I always hoped to wear a set of those, but, unfortunately, it was not to be. Although I spent a great deal of time working in the comp shop, I was never a fully fledged member of that department.

The first thing I did when I started in the experimental department, was make sure that Phil Weaver knew I was the electrical development engineer, a title that as yet meant very little. I assured him that I would be available for any electrical work that his department might require. "Well, actually, we do have a job to start as soon as we receive the car from production. It will need to be prepared for circuit racing," he replied.

I asked him for a specification of the modifications that would be required, and I was quite shocked when he told me that they did not have one. He would give me list of items to be modified or added, and after that it was up to me! After a short bout of panic, I rang Burt Tattersall who told me very bluntly to do what I thought best. It would then be up to the competition department to authorise the changes. I was to give Phil a list of any special components I might need, so that they could be ordered. The car in question, when it turned up, was a pearl grey 3.4-litre Mk I saloon, and Phil's list of changes was relatively short:

1. Move the battery to the boot. (Lucas non-spill type)
2. Fit and wire a changeover switch for a second fuel pump. (SU)
3. Fit and wire a cut-out switch for the automatic choke.
4. Provide means of turning the engine over from the driver's seat, without ignition or fuel.
5. Fit high-rated wiper motor. (Lucas)
6. Supply high-rated dynamo and control box. (Lucas)
7. Supply high-rated starter motor. (Lucas)
8. Fit but do not wire an emergency ignition coil. (Lucas)
9. Supply and fit approved speedometer and rev counter. (Smiths)
10. Supply and fit approved 100psi oil pressure gauge. (Smiths)
11. Disconnect water system from heater matrix (leave fan operating).
12. Fit and wire a headlamp flashing switch.

Having received this list, I then wrote out a specification for engineering to approve, and a list of components that would need to be ordered. I was familiar with the Lucas non-spill battery as not only was it standard on the D-Type, but they had also been fitted on the Meteor NF11, which I had worked on in the RAF. They were also in stock at Browns Lane.

I gave Burt Tattersall a list of the other components I needed, so that he could order them. I discovered that the service department stores had long, ready-cut lengths of heavy duty battery cable already in stock. I never fully understood why these were always in stock, but they suited my needs admirably, and only needed to be cut to the correct length and given the required termination. Once I had everything I needed, I commenced work on my first experimental project.

The Mk I (like the Mk II) had a raised shelf in the front of the boot, to give clearance for the rear axle movement. In the centre of this raised shelf, I placed the battery tray, which I had made myself from folded 14-gauge aluminium. There was a drain hole in both the tray, and the shelf it sat on. This tray was bolted to the shelf with four countersunk set screws, and the battery was held in place by two wing nuts through the lifting handles. The positive side of the battery was earthed to the body of the car, and the negative side connected to the supply cable, which I routed through the boot floor via a grommet, clipped to the inside of the transmission tunnel right up to the front of the body, and then through another grommet in the transmission tunnel. I had worked out that the easiest way to provide for direct turning over of the engine from inside the vehicle, was to mount the starter solenoid on this tunnel. Then the operator could simply press the solenoid's rubber cover to make direct contact to the starter motor. The cable to the starter motor then ran back through the bulkhead, together with the main feed to the other electrics.

The second fuel pump required an additional cable in the body harness that ran to the rear of the car, and a two-way switch mounted on the right-hand dash panel. This panel also held the headlamp flasher switch, which I wired in, via a relay, to the fuse side of the main beam and the ignition switch. A twin cable to the feed side of the choke solenoid (or automatic starting device), and back via a switch mounted on the crash rail, would enable the driver to switch off the solenoid after starting. The rest of the modifications required only the supply and fit of the relevant bought out components.

Whilst I continued to make progress on this car, it was not the only project I had going. Another one involved the first fixed head coupé (FHC) E-Type. I cannot recall where this vehicle lay in relation to everything else, but it was significant to me for a particularly hilarious reason.

Lucas had become insistent that Jaguar should use its new 'Lucar' connector in production, and Bill Haynes, the chief engineer, was keen to save weight (and money). To this end, Lucas had built a number of special electrical harnesses using aluminium-cored, plastic-covered cable, with its Lucars crimped on to the connection points. The harnesses were also in flat strip form, which meant that they could run behind a trim panel, such as the inner sill, without interfering with the trim. There was a significant amount of weight lost, with no additional braiding for protection, and the use of aluminium cores over heavier copper ones, plus the reduced cost of manufacture. Ron and I installed a complete system on this first E-Type FHC as it was assembled.

On completion, everything worked fine, and Norman Dewis put a great many miles on the car while in its initial testing. There were no electrical problems, until that is, it came to pave testing at MIRA.

Those familiar with Belgian pave will know that very few road surfaces have the same destructive capability. Sure enough, on the first run, halfway through the test procedure, the car expired with severe electrical problems, and had to be trailered back to Browns Lane. On investigation, I found that a great number of the Lucar connectors had failed, fatigued at the point where they were crimped to the aluminium core, thus breaking the circuits. I sent a report of my findings to engineering, and eventually it found its way to Rist's Wires and Cables (then a subsidiary of Lucas, who produced the harnesses). From then on, no more was heard of the experiment, and we went back to our traditional wiring methods.

This meant that we now had to build a complete set of harnesses to make the car roadworthy again. With no drawings to refer to, we had to measure for every cable length and connection breakout, produce a dimensional sketch, and build the harnesses by hand, which was a time-consuming operation. We had barely completed all of this, or finished rewiring the car, when engineering asked us to provide the harnesses so that they could be measured and drawn. You can imagine my reaction to that! Fortunately, as it happened, proper drawings could be produced from our dimensional sketches, and supplied to Rist's for the first batch of harnesses to be manufactured.

Further into 1957, Ron Beaty and I were working away in our little department, trying not to be intimidated by the thunderous roar from the engine test cell, where Jim Eastick was running a full-load endurance test on a 3-litre engine. The engine was hooked up to a Heenan & Froude dynamometer, which provided the load, and we had almost become used to the ear-shattering noise. Suddenly, there was an almighty bang, followed by a great deal of clattering, and the noise stopped. For a few seconds, Ron and I looked at each other before he said, "The engine has let go."

For a short time there was absolute chaos. The engine had let go in a big way. One connecting rod had punched a hole through the wooden door and the remains were found some distance away.

We discovered later that Bill Wilkinson had entered the test cell to ask Jim a question. To make himself heard, he had closed what he thought to be the hand throttle control. Unfortunately, it was actually the distributor control, which allowed for degrees of advance or retard to be changed during a run. Suddenly faced with full advance, the engine had tried to tie itself in a knot, resulting in the commotion. We all rushed out to the main shop, half expecting to see serious damage, and, worse still, casualties. I bumped into a very white-faced Jim, who had luckily escaped injury. An incredible amount of energy is produced by an engine rotating at about 6000rpm, and it has to be dissipated somewhere.

As is well-known, the Jaguar version of the 3-litre engine, which was needed following the change in the Le Mans regulations, was never successful. It was

Dave Fielden was a student apprentice from the same intake as Andrew Whyte. After his apprenticeship, he stayed on at Jaguar in the engineering department. He rose to be director of chassis design. He built a couple of very worthy specials in his time; this is one of them. (Author's collection)

unreliable and lacked power. At one stage, very expensive Titanium con rods were tried without success, and I remember seeing Frank Rainbow putting the remainder of them in a vice, twisting them out of shape with a long crowbar, so that if they escaped the factory (which many things did in those days) they could not be used.

I soon became familiar with the experimental department's internal transport, which we all used to fetch and carry components from around the plant. How or why Jaguar came to own a Citroën 2CV, I never knew, but we all made the most of it. We would compete for the lap record around the perimeter road, always aware we might find ourselves on the end of a forklift. The handling was incredible, and it had very soft springing, which often felt like being on a yacht in a heavy sea; the 2CV racing round a corner with the hinged windows flying open was a typical sight. It would not last though, as Lofty England took great exception to this car hurtling past the service department offices, and made sure to put a stop to it.

Another unusual vehicle for Jaguar to own was a pale blue Peugeot 404 saloon, which was also used for external transport. It was quite a pleasant vehicle to drive, and served for many years as the department 'hack.' Another vehicle that came into our hands was a Chevrolet Corvair. This curious American saloon, with its flat air-cooled engine, also had a petrol-fired heating system. I cannot think of anything more unsafe than running around in a car containing a built in fire! Our favourite however, was a Mercedes 300SL gull-wing coupé. Finished in typical Mercedes silver, with tartan upholstery, it went like the clappers. I managed to sneak a drive in it several times on the pretence of evaluating something electrical, and was really impressed with the performance and the handling.

My everyday transport, however, was still the little MG, which was fine in the summer. Without a hood and no heater, though, it was not suited for regular winter driving. Having said that, it had served me well, only failing on two occasions.

The first was when the bearings ran in the vertically mounted dynamo, which also drove the overhead camshaft through a bevel gear. I solved this problem by replacing the dynamo with the shaft from a shock absorber, and mounting a normal dynamo on the offside of the engine, which also allowed me to go from 6 volts to 12, a distinct advantage.

The second issue was more permanent as the clutch started to slip due to oil contamination from the leaking rear crank seal. This was temporarily solved by carrying a CTC fire extinguisher, and spraying regular amounts of the fluid through the inspection hole in the bell housing. This definitely wasn't a long-term solution, and the owner of the local garage was pestering me to sell him the car; that is how I parted with the first love of my life. She changed hands for £75.00, which made me a distinct profit, although I was somewhat miffed to later learn that he had put in a new clutch plate, replaced the oil seal, and sold it on for £175.00. That was a lesson learned.

My replacement for the MG was a very tired Ford 100E Thames van, finished in dark blue, primer, and rust, but at least it ran well and kept me dry. This would eventually transform into a very quick and successful means of transport and competition.

With the addition of a Willment overhead inlet valve conversion to the head, twin downdraft Stromberg carburettors, a four-branch exhaust manifold, and a Murray overdrive giving six gears, it was finally timed through the lights at MIRA at 112mph! Early Ford 105E front discs, front and rear roll bars, and anti tramp rods gave it braking and handling to suit. The addition of side windows, a welded shut rear door, and a pair of XK140 seats in the back also made it a genuine and comfortable four seater. A pair of C-Type bucket seats, carpet throughout, a complete set of instruments, including a Halda Speedpilot, and I was ready for rallying! The final pièce de résistance was a complete repaint in Jaguar Carmen Red, with a grey roof and side flash.

I hasten to point out that those materials and components I sourced from the Jaguar scrap department, were all bought legitimately. Lofty England, however, did not think so.

One day I was called into Bob Knight's office, together with Bill Cassidy, and was asked to explain why my Ford Thames van, parked in the experimental car park, appeared to be more Jaguar than Ford! Fortunately, I had kept the receipts for all my purchases and was thus able to explain my problems away. It turned out that Lofty, on one of his prowls, had seen the van and raised the question with Bob Knight, who then had no option but to follow it through. Bob Knight was never very good at management and was obviously quite embarrassed by the whole episode. Bill Cassidy just dug me in the ribs and said "Well, we got away with that then, didn't we?"

My Ford 100E van after repainting, behind the experimental department. (Author's collection)

Even funnier, some time later, during a Jaguar Apprentices Motor Club (JAMC) driving test at the Daimler factory (where I claimed first place), Lofty came over to me and apologised!

One of the cars that we looked after in experimental was Bill Haynes' personal transport at the time. TVC 420 was a very quick Mk II, with aluminium doors, bonnet, and boot, and an uprated 3.8-litre engine. In one moment of infamy, the car was stolen whilst Bill was at the Earls Court show, and was used as the getaway car in an attempted bank robbery. The police later returned it to Jaguar with no damage, and a chauffeur's hat on the rear parcel shelf. Obviously, the robbers had a sense of humour!

My first contact with the vehicle was when Bill Nicholson asked me to accompany him on a test drive, as there was a complaint from Mr Haynes that the overdrive was not working properly.

Bill, or 'Nicko' as we called him, was a very fast driver, and just a little erratic. He seemed to believe that driving was about going flat out at all times. We set off down the Coventry bypass (not speed restricted back then), and he was quickly up to 80-90mph, flicking the overdrive in and out, whilst negotiating, and swearing at, normal drivers. I tried not to look when he passed on the inside of two cars that did not get out of his way quickly enough – totally barmy!

He tore around the London Road island in second gear, at such a great speed that the inside rear tyre separated from the rim, and the car ground to a halt on three and a half wheels. To make matters worse, TVC 420, for some reason, did not carry a spare, which was probably just as well; guess who would have been changing it in the middle of the A45 traffic? This was also before the days of mobile phones, so Nicko decided that we would have to return to the factory on the rim, not that this lessened his speed much, or improved his Irish temper. When we finally reached the factory, he was straight on the telephone to Dunlop complaining that one of their special test tyres had failed on Mr Haynes' car, and demanding to know what they going to do about it. When I mentioned that we had not concluded the test on the overdrive he remarked, "Oh, don't worry about that, the important thing was that we found out about the faulty tyre." I had to give him credit for his quick thinking if not his driving! TVC 420 has happily survived, and is still, as I write, the treasured possession of a Jaguar enthusiast.

Nicko seemed to be blessed with a charmed life, but he was often in trouble. On another occasion he came to work on his BSA 500 trials bike, and, since he was never one to ignore a challenge, was persuaded to demonstrate his abilities.

Just inside the roller shutter door of the main shop was the hydraulic ramp. This was set up in such a way that it could pivot to allow vehicles to be pushed on to either of the fixed ramps in front of it, so, in theory we could work on three vehicles at the same time, although the vehicle on the hydraulic ramp had to remain mobile so that the other ramps could still be cleared; complicated, but it worked. Nicko moved the unoccupied hydraulic ramp so that it was raised halfway, and lined it up with the first of the fixed ramps (also unoccupied). He then started up the bike about 30ft outside the shop, accelerated, tore through the door, jumped on to the hydraulic ramp, along, jumped on to the fixed ramp, drove to the end, did a 90 degree turn, and repeated this feat in reverse; it was very impressive!

On his way out, though, he just missed Sir William Lyons, who was walking through the door. There was a scurrying as everyone disappeared from the scene of impending disaster. But William did not bat an eyelid or falter in his step. As Nicko, horrified, screeched to a halt, William carried on and disappeared into Bob Knight's office. After a short while, both he and Bob came out and proceeded to inspect one of the prototype vehicles. Sir William left, still with no reaction, and we all thought that Nicko had got away with it again. I had never before, or since, seen Bob Knight lose his temper. Normally, he was a very placid man, but now he was white with anger, as he called Nicko into his office. We never did find out what went on in there, but that was the last time that Nicko came to work on his bike.

Much later, when we had moved to the GEC site, Nicko drove a Sunbeam Alpine. Like most of Nicko's vehicles, it was meticulously prepared, finished in primrose paint, with a black hardtop; it was a very attractive vehicle.

One day, Nicko was late for work, and when he finally turned up he was very much worse for wear. His face looked as though he had been beaten up, and he was covered with cuts and bruises.

Now, being Irish, Bill Nicholson was good with the blarney, and when asked what had happened he described in colourful terms how he had looked out of the bedroom window that morning, having heard a car on the drive, just in time to see his Alpine backing out on to the road, obviously being stolen. In the rush to intervene, he had fallen down the stairs, hence the cuts and bruises. He'd reported the theft to the police, and finally found a lift to work. Sure enough, a little later, he had a call from Warwickshire police to say that his car had been recovered, but it was badly damaged.

All very straightforward it would seem. However, as we learned sometime later, it was all a con. The real story was that Nicko had left for work in his car and, being a little late, was driving at 11 tenths instead of his usual ten. He came haring round a blind bend to find out too late that the road had been dug up overnight, and was coned off with the small paraffin lamps that road repairers used at the time. Unable to take avoiding action, Nicko, the Alpine, and the lamps ended up in a heap in the hole in the road. Being quick-witted, and despite his personal injuries, he quickly worked out that this would not look good, particularly with his insurance company, with whom he was already not on the best of terms. Leaving the car embedded in the hole he thumbed a lift home and then phoned the police to say that his car had been stolen.

The upshot of all this was that the car was written off by the insurance company, but Bill bought it back for a song and rebuilt it to an even better standard than before.

Another larger-than-life character was Stanley Woods, who I became very good friends with. Stan was a big, bluff Brummie. He was also the SU carburettor representative for Jaguar, and what he didn't know about tuning SUs could fit on a postage stamp. He was also a well-known motorbike racer, having competed several times in the TT and other international bike races before the war. He did not suffer fools lightly, but was very approachable, and always ready to help. Stan was never separated from his wooden box of needles and jets, and to see him set up a three carburettor XK engine was like watching an artist at work. Mind you, he did grumble a bit, and if anyone messed with his settings, he would throw a real wobbly, but still, he was highly respected and is a much missed character.

Away from work, much of my spare time was spent preparing the van for competition, as I was getting more and more involved with the Jaguar Apprentices Motor Club. It would be some time before the van was completely finished and ready to compete because it also had to serve as my everyday transport. Every year, the JAMC held an event called, for a reason I still don't know, the Silver Owl Trophy Rally. This was quite a tough timed event, before limitations were put on rallying, and was held in January 1958 on a predetermined course in North Wales, including the Long Mynd. It was an all-night event, and relied heavily on good timekeeping and ordinance survey navigation. This year it started and finished at a garage on the A5, just north of Shrewsbury, and went straight on to white roads. I had persuaded Sam Bacon (another apprentice who would later end up in the competition shop) to navigate for me, and as there was a team prize, we combined with John Austin (in his mother's MG), and Tony Archer (Archers Garage, in Solihull, were Jaguar distributors). We met in the main car park in Browns Lane, before travelling in convoy to the starting line.

To introduce some extra competition, the organisers had invited the Jaguar Drivers Club (JDC) to participate in the event. The JDC members duly arrived and we looked on in envy at the shiny new Jaguars complete with heaters, and their owners wearing fur lined coats and string backed gloves.

We had a motley assortment of vehicles (mostly, I suspect, borrowed from unwitting parents), and were thus in some awe of the competition. As it happened, we need not have worried. The terrible weather meant the roads were atrocious, and it had been snowing hard since we arrived. The Jaguars were not the least bit suited to that kind of weather. I remember hounding an XK150 over the Long Mynd, and watching the sparks as he bottomed his exhaust system until it finally fell off. A little later, I saw a Mk VII disappear off the road and down a long, steep slope like a toboggan. That will take some retrieving, I thought.

At the finish, there were far fewer vehicles, and everyone was tired. No one was clean, but we had all of the control point stamps despite being down on time. When the results were announced, Sam and I were surprised to find that we were first overall, and with a first, second, and fourth finish, we had also won the team prize, all of which made us very happy. The JDC, however, never took part in our night rallies again.

Lofty England was a great supporter of the JAMC, and did much to encourage his apprentices in competition, even to the extent, on this rally at least, of manning one of the control points. You would not find many company directors doing things like that!

I competed in another interesting event with the van. Sam Bacon and I had entered a night rally organised by the Coventry and Warwickshire Motor Club and the Godiva Car Club, both well-known and respected bodies. The event took place in the Peak District on a particularly foul night of rain and freezing fog. The roads, at times, were like glass. I was pressing on through the fog, trying to convince myself that I could see where I was going, when Sam suddenly called out, "T junction, 100 yards, hard right," just as the junction loomed out of the murk. The brakes did nothing to slow us, and the steering had no effect as we slid straight across the junction, and, with a crash, embedded ourselves in the drystone wall on the opposite side of the road. The wall fell down either side of us, stretching 12ft each way, and in the deathly silence that followed, all that could be heard was the bleat of startled sheep who had been hunkered down behind it. Sam had banged his knee on the dash panel, and was howling with what I thought was pain, until I realised what he was saying.

"Stan Pateman is only two minutes behind us in the TR, let's get out." I needed no more encouragement. Stan was a former works driver for Standard Triumph, and had a very quick TR3. The thought of him repeating our mistake and ending up in the back of the van was enough to get us moving. Sam and I, still collecting our wits, stood to one side, and were spellbound as the TR suddenly loomed out of the fog, travelling sideways, and with a loud clatter, having just missed the van, slammed into the wall beside us, demolishing another large section in the process. Very soon we had a small collection of vehicles at the junction, by which point, we sent someone back up the road to warn other competitors, so that no one else joined us in the wall. Remarkably, after some basic repairs, just enough

The badge of the Jaguar Apprentices Car Club, formerly the Jaguar Apprentices Motor Club. (Author's collection)

to get going again, both my van and the TR were back on the road, and we made it to the finish of the rally, although well out of contention.

I have another story of a JAMC rally, which happened two years later, when I was running an interim car, while my van was off road for modifications. This interim car was an early Ford Popular, an E93A. The black paintwork was quite good in some areas, but in others it was reduced to the more usual shades of red rust. Mechanically, it was okay, and it performed reasonably well, particularly in production car trials, where its large diameter wheels, high ground clearance, and lack of wheel-spin inducing power were all an advantage. As a rally car, there was a lot of room for improvement, but we did our best with it!

I remember competing in the Silver Owl Trophy of 1960. The event was held on February 21, and started from Weston Rhyn, near Chirk. I had apprentices Ray Kenney and Peter Wilson, as the all-important navigators. Peter sat in the back of the car, plotting the route for each stage before passing it through to Ray, who would then give me the directions. The weather was atrocious, with very deep snow in places. We came across one group of bedraggled competitors at the foot of a long, steep slope, which had, despite their efforts, defeated all of them. We were told that it was useless to try to climb it. Not being one to be beaten, I decided to have a go anyway, and surprised myself when the Popular, with its tall, thin tyres, and very low power, chugged straight to the top without hesitation.

For a while, we travelled in the ruts, frozen and built up with snow, that had been left behind by other vehicles. They were quite deep but I was able to follow them at a reasonable speed, until they suddenly turned left through a gateway and into a field. The sudden change of direction was too much for the top heavy Popular to follow, and we did a graceful slow roll to the right, ending up on the roof and spinning on the icy road. Orange peel, empty cigarette packets, sandwiches, maps, pencils, and other debris were deposited, like the three of us, on the inside of the roof, as it suddenly became the floor.

Luckily, apart from bruising and dishevelment, none of us were hurt, and we quickly scrambled out through the windows and set the vehicle right. Amazingly, the engine was still running, and did not seem to have suffered from its sudden inversion. We were thus able to carry on and finish the event, although by now, well out of time. It was only afterwards that we realised that we had been following the tracks of a farm tractor, which was why they had suddenly turned into a field. If we had been in any other vehicle than the high riding E93A we never would have got that far. I only had the Ford for about six months, before it made way for something far more competitive.

Finished! The XKSS replica that I built, outside at last. (Author's collection)

The Jaguar E-Type of 1964. (Author's collection)

Above: The Lindner/Nocker lightweight E-Type, after its rebuild. Peter Wilson, who took part in the original build of the car, is standing on the left. (Author's collection)

Left: Me in my Ford 100E van after conversion, outside the experimental department at Jaguar. (Author's collection)

Me holding one of my trophies, the Mike Hawthorn, a replica of his D-Type's 4-spoke steering wheel. This is presented annually by the Jaguar Apprentices Motor Club to the driver who garnered the most number of points at their events over the season. (Author's collection)

My X300 from 1995. One of my favourite Jaguars, named Lady. (Author's collection)

A 1951 Jaguar XK120 OTS, the same colour as the first one that I ever saw in 1949. (Author's collection)

PICTURE GALLERY 97

The Jaguar XK engine that I built to go in my replica XKSS. (Author's collection)

Another view of my replica XKSS. (Author's collection)

Me enjoying a light-hearted moment with old friend and colleague, the renowned Jaguar test driver Norman Dewis. (Author's collection)

The XKSS taking part in the Goodwood tour. (Author's collection)

PICTURE GALLERY 99

The XK120 FHC, which I admired so much. (Author's collection)

*Above left: Norman Dewis testing an early E-Type at MIRA.
(Courtesy Jaguar Daimler Heritage Trust)
Above right: E2A from the rear. (Author's collection)*

Right: Portrait of the author. (Author's collection)

A view rarely seen: the rear of Ian Appleyard's Alpine Rally-winning XK120, NUB 120. (Author's collection)

I drive off in my replica XKSS. (Author's collection)

Chapter 10

1958-1959

The Mk II in competition, E1A, and E2A

It was early 1958 and work was now progressing on what would turn out to be the replacement for the Mk I saloon. In terms of engineering, the changes were minimal. The underpinnings of the car were virtually identical. Disc brakes had been optional on the Mk I 3.4-litre, but were now going to be standard across the range. Power steering and improved cooling would follow.

The rear axle, which had a narrower track than the front, was discarded, and the body was expanded to make room for a wider axle which would make the wheel track the same at both the front and back of the vehicle. The main changes were in the body structure. Gone were the full-framed doors (in my opinion a backward step), to be replaced with chromed side window frames, similar to the big saloons. A larger backlight improved the amount of natural daylight in the car. As these changes were mainly related to bodywork, engineering development was concentrated on the major changes: brakes, cooling, electrical, etc.

Electrical systems were becoming more complex, and the days of the dynamo were numbered. We were beginning to learn about how voltage dropped when travelling along lengths of cable, and so new cable sizes were introduced to help reduced this, which increased the size and weight of the harnesses. As more systems were added, colour coding was modified to suit, and we soon ran out of colours, which lead to colour tracers.

Triplex's early heated rear windows were notoriously unreliable, leading to

A Mk I Saloon. (Author's collection)

much in-house testing before we could justify putting them into production. One of the test rigs that I had to build, was to test sample backlights supplied by Triplex. The sample backlight was put in a long box, to simulate the in-car conditions, and connected to a continuous DC supply at 13.5 volts. Water of varying temperatures was intermittently sprayed, via a pressure pump, on to the outside of the box to simulate changing weather conditions, and the whole test was repeated to produce an expected life span.

It amazes me now, to think how crude these tests were, but the very fact that we had so many failures justified them at the time. Following our continual rejection of the samples supplied, Triplex eventually came up with a seemingly reliable heated glass that went into production. Still, they aged, which meant that not many heated backlights would live as long as the vehicle.

We were also making progress with Joseph Lucas, working together on the replacement for the dynamo – the alternator – and a change to negative earth systems. Today, the alternator is just another piece of electrical equipment fitted on all modern cars, but before that, its development, and the subsequent replacement of the dynamo, was extraordinary. Lights no longer dimmed as more load was applied, but the notoriously unreliable voltage control unit struggled to keep up. In its place went a single unit, which would respond to ever increasing loads at very low turning speeds.

The Mk I 3.4 had already made a name for itself in saloon car racing, and the Mk II would continue its success. Before being overpowered by more advanced machinery, the Mk II ruled the race tracks of the world.

The really successful cars were produced in the competition department at Browns Lane. People now postulate about the preparation and success of cars that ran under different banners, such as the John Coombs cars, which now command increasingly high values, but none of the preparation work was better than that of the factory. High lift cams, gas flowed heads, cut-away spats, free-flow exhaust systems, simplified inlet tracts, reinforced rear axle and Panhard rod mounts, Koni shock absorbers, and improved braking had all been tried, tested, and proved by the competition department. It did not matter to Jaguar who claimed the kudos for preparing the car; what mattered was that it won, and it did, over and over again. Developing the Mk II into one of the most successful saloon car racers of all time was an ongoing process.

The publicity of international rallying was increasing, the importance of which was not ignored. Ever since the successes of Ian Appleyard and Pat Lyons in the works XK120, the NUB 120 in the Alpine trials, Ronnie Adams in the Mk VII at the Monte Carlo Rally, and the Mk VII and Mk I in saloon car racing, Bill Lyons had been well aware of the effect that such achievements would have on general sales.

Now that it was relieved from the frenetic preparation of works cars for international races such as Le Mans, the competition department threw all of its efforts into saloon car racing and rallying, with the occasional privately owned C or D-Type thrown in for good measure, depending on the status of the owner!

There are two works prepared Mk IIs that still stand out in my memory. The first was prepared for the Whitehead brothers, to run in the 1958 Tour de France, and the second for Bernard Consten, run in later years, in the same event. One of the major conversions for both cars was the installation of long range fuel tanks. The additional tank occupied the spare wheel compartment, and was linked to the standard tank so that both were filled at the same time. The spare wheel now sat on top of the compartment's cover, which was pop riveted around its flange. The auxiliary tank was easy to remove, if necessary, by drilling out the pop rivets.

This modification lead to a quite hilarious situation when it was carried out on the Whitehead car. Joe Sutton had installed the auxiliary tank, riveted down the spare wheel cover, and fitted the spare wheel. The time came for a leak check, and a careful recording of the actual capacity of the combined tanks. Joe was busy filling up from the hand-operated bowser that the competition shop used, when there was a loud cry of, "fuel leak!" Joe looked down to find himself standing in a growing pool of petrol, which was pouring out from under the rear of the car. There was a lot of scrabbling around to find containers in which to catch the forty odd gallons of fuel that was leaking from the car. After the panic had died down, all of the connections were found to be fuel tight, and it seemed the fuel had leaked from the top of the tank, which confused us all. The only thing left to do was take out the rivets and remove the spare wheel cover. Having done so, the problem became obvious. When Joe had drilled to put the rivets in, he had also drilled through the upper surface of the tank. Result: one scrap tank.

The tank was saved by Bob Blake and his excellent handling of metal. He carefully drained every last drop of fuel, and filled the tank to the brim with soapy water. He then soldered up all of the holes created by the wayward drill. We all stood back with hands over our ears, fire extinguishers at the ready, which was totally unnecessary as Bob knew exactly what he was doing.

The rest of the electrical specifications on that car were identical to the cars raced on the circuits. The only additional modifications that I recall were to remove the centre console and trim the gearbox cover with carpet (similar to the Mk I), and fit two neatly trimmed bucket seats (similar to the C-Type). I installed a Halda Speedpilot in the glove box, and connected it to the speedometer with a double-ended angle drive, supplied by Smiths for that purpose.

I had great fun calibrating the Halda, as this could only be done on the road over a precisely measured distance. Conveniently, Warwickshire County Council had installed two marker posts on the A45, just below the old Humber factory, and the distance between them was exactly 1760 yards. On this stretch of road, I was able to trip both the Halda and the speedometer at the first post, drive to the second, and check the recordings. Both were unacceptably inaccurate. The Halda could be adjusted, and after several further tests it was within the acceptable limits. However, the speedometer could not be reset as it had been constructed by Smiths to suit the axle ratio, which if memory serves me correctly was 3.77:1.

This mystified me, as I knew engineering would have given Smiths all the data (axle ratio and tyre size) it required in order to produce an accurate instrument. After I returned to the factory and had checked the specification, I discovered that they had been given the size for standard production tyres, which had a different rolling radius to those fitted on the car.

We were running out of time, so Smiths was called with the correct information for a replacement to be made. Three days later, when I was still fitting the new speedometer, Norman Dewis arrived, needing to carry out a shakedown run. That is how I ended up doing the run with him, and he was not too happy at having to drive slowly between the two posts on the A45. Still, I had the pleasure of seeing Norman at work, and he used me to record the list of snags that would require rectification before the car would be released.

The worst of these snags was the grabbing rear brakes, which were cured by a pad change, and the car's tendency to wander when travelling at high speeds, which was solved by an adjustment to the Panhard rod, which located the rear axle. I had also installed Lucas Le Mans headlamps with 100/100 watt bulbs, and a pair of 'Flamethrower' spot lamps, connected to the main beam circuit.

The Consten car, in contrast, presented more of a problem. It had been decided that its 3.8-litre engine would have three SU HD8 carburettors, basically an XK150 S setup. This created a major problem both for me and the engineers.

Firstly, the clutch master cylinder had to be changed to horizontal operation in order for it to clear the rear carburettor, and the wiper motor had to be removed, so that a cut-out could be made in the offside inner wing valence in order to clear the front carburettor. My problem, then, was where to put the wiper motor. I solved this by eventually mounting it on the back of the loose plate in the driver's kick panel, at the foot of the 'A' post, which would normally house the right-hand speaker for the radio. I then had to shorten the rack drive to the wiper motor, and make up a 'bundy' drive tube to run inside the car, up and over the right-hand instrument cluster to the first wheel box. All of which worked exceptionally well. Many years later, when I had the opportunity to inspect this car (now with a new owner), I was very pleased to find my original wiper motor and assembly still in place and still working.

We finished building one very special car in April 1958. "Well, actually," as Phil Weaver would say, we started to build it late in 1957, but it was six months before serious testing began.

E1A (A for aluminium) was conceived by Bill Haynes and his engineering team, as early as 1954, mainly as a suspension test vehicle. It was generally recognised that the days of the live rear axle on competition cars were numbered, and a replacement for the XK150 was a priority.

Both Sir William and Malcolm Sayer were of the opinion that the shape of the long-nosed D-Type had a natural progression, and that was the way to go, both in terms of aesthetic appeal and aerodynamics; how right they were.

In later years, after his retirement, Tom Jones, a senior engineer in chassis design, confirmed that the necessity to get away from the traditional solid axle had been known by the engineering department for some time. The problem they had was that it would have been very difficult and expensive to re-engineer the XK150, because of its separate chassis. The unitary construction of the E-Type later solved that problem.

De Dion was an already tried and tested system, and had proved its worth on some competition cars. In fact, a D-Type had been converted and tested with such a system, but the results were inconclusive, and the system did not lend itself to a production sports car. Serious consideration had been given to a system that used unequal length swing axles (similar to that already in production at Mercedes-Benz),

*One of only two poor quality photos taken of E1A at Patrick Jenning's house in North Wales.
(Courtesy Jaguar Daimler Heritage Trust)*

and indeed a 190 SL saloon had been evaluated by Jaguar just for this purpose; it was, however, also discounted.

The final conclusions of the team would up end dictating how all Jaguar rear suspensions would operate right up to the Series 3 XJ, and, with modifications, these suspensions would continue to appear in the current range. The downside of the new suspensions was the increase in weight. To eliminate the amount of noise and vibration that could be heard and felt whilst in the car, the differential would need to be mounted in an insulated steel cage. The main stress member, the lower wishbone, would have to be very substantial to cope with torque, and trailing arms would be necessary to limit axle wind up. This was the beginnings of the system that would end up on E1A. The vehicle itself was assembled very similarly to the D-Type, with a riveted and welded aluminium monocoque, and a separate chassis sub frame to carry the engine and transmission.

For reasons I have never been able to fathom, it was decided that the 'little green car,' as Phil Weaver called it, would be a scaled-down version of its eventual sibling: the E-Type. Because of this, and almost as an after thought, a 2.4-litre power unit was used. If cost was being considered, I'm sure I'm not alone in wondering whether, for a one-off build, producing a full-size vehicle would have been that much more expensive. The 3.4-litre power unit was by far the obvious choice! If all they wanted was a 'test mule,' why build a special little car? Could it be that Jaguar was seriously considering a small sports car at the same time?

Mechanically, as far as I can remember, the front suspension was the same as that on the XK150, with the track adjusted to suit the car's dimensions. The Salisbury differential was bolted directly to the body, and drove through equal length driveshafts. The lower wishbones were fabricated and boxed, and tubular, adjustable trailing arms were rose jointed, both to a body bracket and the wishbones. The brakes were Dunlop discs, with detachable pistons operating through twin master cylinders with an adjustable balance bar and no servo. I drove E1A on several occasions, mainly under the pretence of checking the electrical systems. It handled well and, despite the 2.4-litre engine running on 2 HD6 SU carburettors, it was surprisingly agile.

E1A was an exquisitely proportioned, if somewhat spartan, vehicle. Its electrical systems were minimal. As it was never intended to run in the dark, no headlights were fitted, only side, tail, and stop lights, to conform to the legislation of the time. There were the basic instruments: the speedo and the rev counter, with a centrally mounted ammeter between them; oil pressure and water temperature gauges on either side; and a single Mk I horn operated by a D-Type push button to the right of the dash. Further switches – connected to the side lights (from the redundant material stores, and once used on the 1.5-litre saloon), standard D-Type tail lights, (ex-Morris Minor), SU fuel pump, and ignition – were fitted at the base of the Perspex windscreen, completing the dashboard. At this point, my main jobs were to make up the minimal harness that was required, construct the instrument panel, then test the systems.

The construction of E1A was highly secretive, and no one really knew the thought-process behind it. The rest of the story is well documented by historians with a much greater knowledge. It is such a pity that no one had the foresight to preserve this beginning, the genesis of what would become probably the world's most successful and beautiful sports car.

In January 1959, we were all shocked to hear of Mike Hawthorn's death. He had died in an early morning crash on the Guildford bypass, while driving his 3.4-litre Mk I saloon. The vehicle inexplicably left the road at high speed, and crashed into one of only a few trees at the road side. The impact on the driver's side almost dissected the car, and Mike had died instantly.

He had been a very popular member of the Jaguar team, and dying in a road accident seemed somewhat ironic. The sad remains of the car were eventually returned to Browns Lane, where the competition department carried out an investigation, in an attempt to shed some light on the cause of the accident.

A manufacturing fault would, of course, have been very bad news for the company, especially with such a high profile driver. However, no mechanical fault could be found, and the eventual conclusion was that it was sheer bad luck, a case of being in the wrong place at the wrong time. Many theories have spread, one saying there was use of a hand throttle which jammed open. The investigation disproved this theory, as no such device was found to be fitted to the car. Eventually, the remains were cut up into small sections, and carefully disposed of to avoid souvenir hunters.

Later in 1959, we started work on the next phase of the E-Type development programme: E2A. Around the factory, this was rumoured to be a direct replacement for the D-Type, and a team of at least three cars would herald Jaguar's return to the race circuits, particularly Le Mans. However, this turned out to be a pipe dream, as only one car would be built. We were always led to believe that Bill Haynes and Lofty England were very keen on the idea of returning to racing. Meanwhile, Sir William would only allow the build of the first car providing it met two conditions. Firstly, that it would play a part in the development of the E-Type road car, and secondly, that there was proof it could compete with other sports racing cars of the day. As things turned out, neither of these conditions were met. The car would race once at Le Mans in 1960, but as a private entry with the Briggs Cunningham team.

To me, the car never looked quite right. With its high Le Mans specification

E2A outside the factory. (Courtesy Jaguar Daimler Heritage Trust)

windscreen and narrow body, it had lost the subtle curves of the D-Type. Its 3-litre aluminium block engine had always been suspect, and although it reached 190mph down the Mulsanne Straight during test weekend, reliability would be its downfall; after 89 laps of the actual race, it had to be pushed away because of a broken engine. All of this seemed to validate Sir William's caution, as he would certainly not have returned to racing without an almost guaranteed win.

The car had run at Le Mans in the Briggs Cunningham colours, white with two blue stripes down the length of the body, and after the race it was shipped to America 'on loan' to Cunningham. There, it was modified quite a bit. The tail fin was removed, leaving a hump over the petrol filler cap, similar to the production D-Type. The Le Mans windscreen was cut down, and replaced with a much shorter Perspex version, which wrapped around to the side screens. Finally, the 3-litre engine was replaced by the much more reliable and powerful 3.8-litre iron block unit. To accommodate the extra height of the new power unit, a bonnet bulge was required, a feature which would be carried over to the production E-Type. All these features are still present on the car to this day.

The car was then raced in the States, with moderate success, but was eventually retired to Cunningham's private collection. From there, it was later returned to Browns Lane, where the service department promptly changed the colour scheme to BRG. It then sat forlornly in a corner of the competition department, covered with dust sheets. At some stage in October 1961, the dust sheets were removed and work began on converting the rear brakes to accommodate the experimental Dunlop 'Maxeret' anti-lock braking system. For those not familiar with this system, which was a forerunner of the modern ABS, the basic principle consisted of a small generator attached to either hub, driven by a toothed belt from the drive shaft. The electrical energy generated would be used to hold open a solenoid valve in the brake line, thus allowing braking to each wheel. The theory was that as soon as a wheel locked, the brake line to that wheel would be bypassed, allowing the wheel to spin again. The system was already in use in the aircraft industry, and had proved itself as an adjunct to safety. However, stopping only one ton of motorcar proved to be very different to stopping several tons of a modern jet aircraft.

My part in all this was to wire in the electrical system for a shot firer. This consisted of a magazine fixed under the frame of the nearside front suspension, with the exit holes pointing to the road surface. If my memory serves me correctly, this magazine held ten .22 cartridges, which, instead of a bullet at the tip, held a chalk block. An isolating switch on the instrument panel fed a micro switch on the brake pedal, via a fuse. When the switch was closed by the pedal, energy was supplied to the first shot in the magazine. A mechanical shuttle in the magazine then selected the next shot to be fired. The theory was that as soon as the brake pedal was depressed, the shot would fire a chalk cap at the road, thus giving a start point. Measurements could then be taken between there and the final stop point, allowing a comparison to be made with and without the Maxeret system.

Eventually, the installation was completed, and we took the car over to Honiley Airfield, near my home in Balsall Common, for testing. Johnny Frost of Dunlop, Derek White (the man in charge of the project), Bob Penney (from the competition shop), and Norman Dewis were all present. We did several runs, and every time there was trouble with the shot firer. The shuttle in the magazine kept getting itself confused, resulting in several shots being fired in rapid succession, until, much too soon, we ran out of bullets. A great deal of inconclusive data was gathered, E2A was returned to its dust sheets, and the Maxeret system died a quiet death.

Chapter 11

1960-1961

The Grand Prix, my first Jaguar, and Sandy Lane

Compared with what had gone before, 1960 was a relatively quiet year. The Mk II was being developed, and one of our Mk I 'hack cars' was now running with independent rear suspension. We were still working on the prototype E-Type.

Ron Beaty had now moved into the experimental engine department, leaving me alone in experimental electrical development. It wasn't long before I was joined by two new recruits: Len Hives and Tony Boden. Tony, like me, came from the production lines, but Len was a new employee who had been working outside the motor industry as an electrician. They were just in time, as work on the E-Type was about to step up a gear, and the prototype of Jaguar's first large saloon car with independent rear suspension (the Mk X) was beginning to take shape in Fred Gardner's design shop.

At this time, all my efforts were concentrated on perfecting the new triple-blade windscreen wiper system for the E-Type. Engineering had decided that three blades were necessary in order to effectively clear the entire E-Type windscreen. This required a complete departure from the normal rack drive through wheel boxes method, as Joseph Lucas was concerned that three wheel boxes would be too much for a rack system to handle. The solution to this problem was to introduce a mechanical linkage across all three wheel boxes, directly driven by an operating rod mounted to a cam arm on the motor. The linkage was mounted inside the bulkhead, behind the instrument panel, and was notoriously difficult to get at.

The motor was mounted on the engine side of the bulkhead. The connection between the control rod and the linkage was a ball and socket, similar to the one already in use on throttle linkages, with small snap ears to lock the connection. The snap ears had a nasty habit of falling off, and it was particularly difficult to reassemble on the finished car, as the only access was from behind the centre drop-down instrument panel. You were also lucky if the control rod did not flail around and short out the wiring to the fuses, which were situated behind the panel on the bulkhead.

I remember one occasion, much later on, which typified this problem. I was in the paddock at Silverstone, possibly for the 1963 *Daily Express* Trophy Meeting, when I was approached by an agitated Lofty England. "Ah, Martin, found you. Come with me," he said in his imperious manner, and strode off with me meekly following, despite the fact that I was there on my time, not Jaguar's. He stopped at a primrose yellow E-Type, and I recognised the rather angry-looking man standing beside it as Innes Ireland. I also did not fail to notice the very attractive blonde lady sitting in the passenger seat, who turned out to be his wife, Jean.

"I have found the man Innes, he will sort you out." With that, Lofty disappeared back to his duties as clerk of the course. Innes looked balefully at me and said, "I understand that you are responsible for the wipers on my new car." Before I could respond, he went on. "I have driven all the way from London in the pouring rain, with no wipers, and the most terrible noise behind the panel. What are you going to do about it?"

The Mk II saloon, which could be bought with either a 2.4-, 3.4- or 3.8-litre engine. (Author's collection)

I have omitted the expletives that came with this one way conversation, but I'm sure you can guess what they were. I knew straight away exactly what had happened; the disconnected arm was thrashing away merrily inside the bulkhead. I explained my suspicions to Innes and told him that I would attempt to reconnect the arm. I said that there was no guarantee that it would stay connected afterwards, as the small clips that locked the connection could be faulty or damaged. "Okay," he said, "I will leave you to it. I have to go and drive a car in the next race. Oh, and try not to look up my wife's skirt if you have to get under the dash," he said with a leer and off he went. So, there I was, lying on my back across the drivers seat, under the steering wheel, with one hand through the access hole in the centre of the bulkhead. I was trying to return the locking clips to the open position, conscious of Jean Ireland's legs, just a few inches away, and concentrating very hard on not looking.

I finally managed to prise the clips open and reattach the arm onto the linkage, but I could see that they had been damaged as the arm had thrashed around, and were never going to close cleanly. Attempting a belt and braces job, I bound the connection with some plastic insulating tape, borrowed from the pits mechanic, and hoped for the best. Having put it all back together, the wipers were now working. I explained to Mrs Ireland that the repair was not permanent, and that the car would have to go back to the dealer for proper attention. She smiled sweetly, and said that it would probably be able to find its own way there it had been back that often. She thought that I was a clever man, as her husband was useless at anything mechanical, and with that I already felt rewarded.

An hour later, I was surprised to hear an announcement over the PA: Brian Martin to report to Race Control. Oh God, I thought, it's fallen off again. When I arrived, Lofty was there with a smiling Innes. "I want to thank you for repairing my car in your own time," he said, holding out a five-pound note, which I declined to take. I pointed out that the repair was only temporary, and that the car should go back to the dealer as soon as possible for a permanent solution. He again thanked me, and as I turned to leave he said, "And what did you think of my wife's legs?" Somewhat nonplussed, I said that they were very nice. "Good," he said.

Some time later, I was participating in a JAMC driving test at the Radford factory, where Lofty was one of the judges. He came over to me and said that he had been remiss in not thanking me before, and that the car had been back to the dealers and had the arm replaced. He was human after all!

Many years after that, I was reminded of this incident, when I took my replica XKSS on the Innes Ireland Memorial Run to Silverstone with Richard Noble. The car behind us was a 250 GTO, driven by Sir Stirling Moss. In the passenger seat was Jean Ireland. At lunch, I reminded her of the problem with the wipers on the E-Type all those years ago, and, bless her, she remembered.

In 1961, the time had come for the experimental department to make its final move. Jaguar had now taken over what has always been referred to as the GEC block, which was outside the area of the main plant. It consisted of a large, two-storey office complex and several outbuildings, one of which became our new home. The old GEC entrance became our south gate, and there was no need for us to enter through the main gate, further up in Browns Lane. The new building was about twice the size of the one that we had just left, which meant the departments were properly separated, by more than just partitions. The main shop was well lit, with natural daylight from the north light glazing, and contained everything needed to maintain the growing fleet of experimental vehicles. We even had a car wash operated by Ted the 'Pole.'

Ted had been a member of the Free Polish Army during the war, and had stayed on in England afterwards. He was a very excitable little character who spoke broken English, and would lapse into his native tongue when stressed, which happened fairly easily. He looked on the car wash as his personal property, and woe betide anyone who interfered with its operation. Barrie Woods was the joker among us, and would regularly wind Ted up. We would often witness Barrie

The prototype Series 1 E-Type OTS. *(Courtesy John Starkey/Jaguar Daimler Heritage Trust)*

My replica XKSS on a rally. (Author's collection)

being chased around the department by Ted, sponge and brush in hand, shouting obscenities in broken English or sometimes in Polish.

We now had a reasonable-sized machine shop, three four poster ramps, a proper welding booth, and Bill Cassidy had his own office. Next to the main shop was the experimental body shop, which was at least twice the size that it had been in the old shop, and next to it, but slightly smaller, was the competition shop. A passage ran the length of the building, and on the other side of that was the engine assembly and test room, the silent room, the electrical department, and experimental stores. Thus began our new life!

The old GEC office block became the new home of Jaguar engineering. This was only across the road from the experimental department, and so we had less warning when senior members of staff were approaching. Sir William would soon perfect a nasty habit of driving to the GEC car park, walking through the drawing offices in the GEC building, nipping across the road, and suddenly appearing at the doorway. The effect of this should not be underestimated; Most of us were in total awe of this man, and it was like a god suddenly presenting himself.

On one hilarious occasion, Barrie was busy winding up Ted. While Ted had been temporarily absent, Barrie had crawled under the grating of the car wash, with the intention of suddenly appearing from the ground when Ted returned. Unfortunately, Barrie had not expected Sir William to come through the roller shutter door to the right of the car wash. We all stood open-mouthed, as Sir William, hand in pocket, looked around before calling Bill Cassidy over. By this point, Barrie was frozen in terror, and Bill disappeared into Bob Knight's office, reappearing a few minutes later with Bob in tow. Sir William and Bob then had an enthusiastic discussion, blissfully unaware of the drama going on below ground. Ted, meanwhile, had returned to his beloved car wash, and, seeing Sir William, was determined to look busy. Turning on the water hose, he proceeded to wash the surrounding floor, not knowing that he was drowning an unfortunate Barrie in the process. Eventually, Sir William, satisfied with his conversation, turned on his heel and walked out.

Not a second later, Barrie had erupted from under the grating, looking like a drowned rat, terrifying Ted in the process. Bob Knight, bless him, just stood there with a smile on his face, rocking on his heels and shaking his head, before he disappeared back into his office. By now, we were all in hysterics, as Ted chased a sodden Barrie. Bill Cassidy, though, was on the warpath, having just received a phone call from Bob Knight. Barrie was very subdued when he later emerged from Bob Knight's office, although that didn't last long. You cannot keep people like Barrie down; and despite his pranks, Barrie was a brilliant engineer.

Back at home, most of the modifications that I had envisaged for the van had been completed and it was going very well, although the Murray overdrive was grossly unreliable. Being a mechanically operated transfer box, its operating lever had a sequential fore-and-aft movement and, invariably, it needed to go in the opposite direction to the gear lever in order to keep the ratios building. For example, the move from overdrive third to top gear, meant pushing the overdrive lever forward, while pulling the gear lever back at the same time, which could be a confusing manoeuvre. It also had the nasty habit of falling apart internally, and suddenly there would be no gears at all! I remember one day, I removed the gearbox three times before I finally had it sorted. There was just one final

Left: Scuderia Centro Sud's Cooper T51 Maserati at Silverstone in 1960, for the British GP that I attended. It was driven by Ian Burgess but failed to finish. (Author's collection)
Right: The Vanwall, which would go on to win the British Grand Prix at Aintree in 1957, being pushed to its pit. (Author's collection.)

temporary modification to make, which almost landed me in trouble with Lofty, yet again.

Barrie Woods, Peter Wilson, and I, together with three more of our pals, had decided that we would go to the British Grand Prix at Silverstone (we had been before, both times to Aintree, once in 1957 and once in 1959). We could not afford grandstand tickets, but at that time, a car and all of its occupants could enter the outer perimeter of the circuit for a single modest fee – the more people in one vehicle, the less it cost individually. We managed to squeeze six people into the van and, because I had provided our transport, I entered for free. We also bought a paddock pass for £1 which we put inside a Swan Vestas matchbox and tossed over the fence repeatedly until all of us were in the paddock. Peter Wilson, however, went one too far when he sneaked into the pits gallery, only to be ejected by the security carrying out a ticket check. It was press day after all.

Even after all that, we still could not see the circuit properly, so for the next event, we decided to build a platform on top of the van. Now, I have to admit that this was done in works time, behind the experimental department. We built a roof rack type frame to cover the entire roof area, which we temporarily attached to the van's gutters. The frame then held a sheet of three-quarter plywood, which became the viewing platform. The final touch was an aluminium ladder to climb up, which I persuaded Bob Blake to weld for us. It worked beautifully, and we watched the racing in style, with a marvellous view on the outside of Becketts corner. We even had the effrontery to wave at Lofty as he completed a circuit as clerk of the course, in the black works XK120. We stopped at The Green Man for a couple of pints on the way back to Coventry.

About a week later, I walked into the main workshop, and there was Lofty, in deep conversation with Phil Weaver. As I went to pass them, he grabbed me by the arm, excused himself from Phil, and pulled me to one side.

"Now then young Martin, I want to talk to you," he said. "As you know, I was officiating at Silverstone last weekend, and as I drove around the course, I was intrigued to see half of the experimental department standing on top of your van at Becketts, waving like lunatics. It struck me that you must have done a pretty good job with that platform for it to carry all of that weight. I have been watching your progress with great interest and curiosity."

By now, I was standing tongue-tied and open-mouthed, waiting for the axe to fall.

"Well, now, perhaps you can get back to working for the company, and show the same amount of ingenuity to the job you're being paid to do," he said, eyebrows bristling.

"Yes, Mr England," was all that I could reply.

"I am glad that you did not deny it. I hope that you enjoyed the racing." He turned and walked away. We never used the platform again, but I did include it as an accessory when I finally sold the van, and Lofty had left yet another indelible mark on me.

RVC 591 had been one of the first 2.4-litre Mk I press cars and was used for several of the launches of that model, after which it joined the experimental fleet as a development car. It was used for several new projects, including the fitting of independent rear suspension, the system that would eventually see production on the E-Type.

Another trial was fitting SU carburettors, and Ron Beaty was in charge of the in-car installation. With Solex carburettors, manual chokes were fitted as standard, and these required no wiring in the forward harness. However, these new carburettors would be fitted with an automatic choke, and it was my job to wire it in. After completing the installation, Ron suggested that we take the car out for a quick road test, and do the final settings on the road.

The problem was although RVC 591 was road registered, it was not taxed, and trade plates would, therefore, be required. Each set of trade plates had to be purchased by the company, and an annual fee paid to maintain them. With Jaguar's tight control of finance, the number of plates we had was kept to an absolute minimum. Only three departments were deemed to require them: the service department, to road test customer vehicles; the production road test department, for new, unregistered vehicles; and the experimental department. As the numbers were so severely limited, it was often necessary to borrow a set of plates from one of the other two departments. If my memory serves me correctly, the experimental department was issued with three sets of plates.

The department's most famous set, of course, was 774 RW, regularly seen in photographs of works C and D-Types at Le Mans, and made immortal on Mike Hawthorn's winning D-Type of 1955. This can still be seen, recorded for posterity, on Nigel Webb's recreation of the very same car.

It turned out that our three sets were already in use, so we called in on road test. All of their plates were in use too, so it would have to be the service department. I decided that I did not want to have another run-in with Lofty, so Ron went and turned up some time later with a set of trade plates under his arm.

With Ron driving, we went out of the south gate into Browns Lane, and headed for the A45 and Meriden, which was our regular test route. All went well until we reached a post house on the crest of a hill. Here, a sharp left turn was required in order to stay on the road. I quickly realised that Ron was struggling, swearing to himself, as we continued in a straight line across the apex of the corner, and ground to a halt at the entrance of the post house, narrowly avoiding a Coventry Corporation bus in the process.

Following a short silence, I asked what had happened, and Ron said that the steering had suddenly locked. Mystified, we opened the bonnet and the problem became obvious. A short T-bar, which we presumed had been left on the inner wing valence, had dropped into the engine compartment, and, as Ron had started to turn, had neatly wedged itself in the knuckle joint of the upper column, thus locking the steering; it was an unusual, and not very funny thing to happen.

After removing the bar, and checking around for other potential problems, we continued with the test in a very subdued manner. The performance and flexibility was much improved with the two HD6 carburettors, and even more so when Stan Woods came in and made some final adjustments. Twin HD6s would later become a standard fit on the 2.4-litre Mk II, replacing the Solex ones.

This modification, however, did bring other problems. The complete inlet manifold and carburettors had to be removed from the engine, before the body could be lowered over it, whereas, before, on the earlier model, the body would pass neatly over the Solex carburettors.

It was sometime during the summer that I had another traumatic experience, this time with the van. One lunchtime, I was returning from a quick dash home to Nuneaton, via the back roads through Fillongly. Passing through Wood End at a fairly high speed, I suddenly spotted something that looked like a piece of metal, lying directly in my path. With no time to take avoiding action, I aimed to pass directly over it.

It all happened very quickly; there was a huge bang, the back wheels locked suddenly, and I skidded to a halt at the side of the road. Having collected my scattered wits and exited the car, the first thing I noticed was a pool of oil quickly collecting under the engine. On further inspection, I found the cause. Jammed tightly between the transmission tunnel and the propshaft, which was now bent at an angle of about 30 degrees, was a shiny ploughshare. This was what I had seen on the road, with its sharp end pointing towards me. Sticking up a little higher than the ground clearance of the lowered van, it had neatly carved a groove in the front suspension cross member, before dissecting the whole length of the sump, and then leaping up to jam itself in the propshaft: disaster!

Luckily, there was a telephone kiosk only 200 yards away, so I made a panicked call to Jaguar and was put through to Bill Cassidy. I explained my dilemma. His initial response was to ask me what the hell I was doing out at Wood End in the lunch break, but then he told me to stay where I was and he would try to get someone out to help. I realised that the van could not be driven, and even if being towed, the offending article would have to be removed. I jacked up the car to remove the propshaft and the ploughshare, which I kept as a souvenir for some time.

About 30 minutes later, Harry Hawkins, driving the works Peugeot, turned up with a tow rope. After the usual leg pulling, which Harry could never resist, I arrived back in Browns Lane on the end of the rope. This had all taken at least two hours of my time, and I was well and truly late. Bill Cassidy, bless him, told me not to clock-in, and signed my card to say that I was away on company business, so at least it did not cost me any pay. After thanking everyone profusely, I had to get back to work. No one would believe what had happened until I pulled out the ploughshare as evidence.

I had to leave the van where it was until repairs could be made, and scrounged a lift back to Nuneaton in the evening with Barrie Woods. At the time he had a Standard Eight, which he drove flat out everywhere. He frightened the life out of me by passing over the halt sign at Fillongly crossroads without even lifting his foot, something which he apparently did on a regular basis. He had a theory that if he went across fast enough, there would not be time for anything to hit him; it was madness! I got lifts with Barrie for three days, while I arranged the repairs for the van. By the end of it, I was a nervous wreck and vowed never to travel with him again.

Having purchased a replacement sump from the local scrapyard, I was given permission to use one of the ramps in the experimental department, after works hours, to carry out the repairs. Harry, meanwhile, came up trumps, and he and Frank Lees heated and straightened the propshaft. Harry even spent his own time welding a plate over the split in the front suspension cross member, and would take no form of remuneration for it, other than a pint at The White Lion. I then fitted the repaired propshaft and the new sump. Later that evening, I drove the van home and it was as though nothing had happened. All it cost me was a few pounds and a lot of dented pride.

Around the same time, I bought my second Jaguar, which was to be another of my mistakes. I was totally besotted with idea of owning an XK120 FHC, and I had found one for sale at an affordable price. I have no idea how this sorry-looking car came to be parked in front of a council house in Bedworth, but I bought it on sight, and parted with a hard earned £75 for the suede green beauty.

I soon realised my mistake, as it died in a cloud of steam on the way back home. This turned out to be no more than a broken fan belt. A local farmer came to my rescue, and towed the vehicle into his yard, before driving me home in his Land Rover. He would not take a penny for his trouble, and even helped me fit the replacement fan belt when I returned the following day.

The XK had been badly neglected, and needed a lot of work, although mechanically it was not too bad. I gave the car a good clean inside and out, and a cut and polish. Now, it looked reasonably presentable, so much so that, only two months later, the man who had bought my MG offered me £175 for it. I could not resist such a profit, so I sold my Jaguar. After the sale of the MG some years before, I now felt that we were even. It was time to start looking for a replacement!

The replacement turned out to be a 1956 2.4-litre Mk I saloon, which was in a good condition and cost me £425. This was a sum of money that I could ill afford, which is exactly what my bank manager told me when we were negotiating the loan I needed in order to complete the purchase. It came from Arden Street Motors, which was owned by a friend of mine, in Earlsdon.

The car was British racing green, with tan trim and no rust, and ran very well. After a cut and polish, the paintwork came up beautifully, and a good clean inside did wonders for it. At last, I owned a motor car that I had been instrumental in the building of.

One of the first things I did was replace the strangling downdraft Solex carburettors, using a 3.4-litre manifold and HD6 carburettors, long before Jaguar did the same thing on the production Mk II. Stan Woods was a great help here, and spent some time setting them up. The addition of an XK120 twin-pipe silencer, coupled to a modified 3.4-litre downpipe system, improved the performance. The next step was to pull the cylinder head off the block, skim off enough to raise the compression ratio, and have Les Ryder gas flow it.

Les is probably better known for his work on the A-series BMC engine cylinder heads, which were used extensively in saloons and early Formula Junior

The prototype Jaguar E-Type in 1959. (Courtesy Jaguar Daimler Heritage Trust)

The production line for the early Jaguar E-Type, Browns Lane, 1961. (Courtesy Jaguar Daimler Heritage Trust)

cars. When he retired from the experimental engine department at Jaguar, he's spent many years doing this, building a reputation for himself, and becoming very successful. All of the work was carried out in the garage of his home in Coventry, and he would spend hours grinding away with rotary arbours and emery strip to achieve the best gas flow possible. Les was an ex-'desert rat,' who was mentioned in despatches for his bravery at El Alamein in the North African campaign of World War II. Just before his death, he took part in a televised return to that battlefield: a very poignant moment for him. Les, and his wife, Chris, became very good friends of the family, and would often visit us in later life.

To complete the work on my Mk I we installed a pair of modified camshafts, which increased the lift, but, more importantly, extended the dwell period. In this form, the 2.4-litre was as quick as a standard 3.4-litre. In retrospect, it would have been easier to just fit a 3.4-litre unit to achieve the same result, but that would have spoiled the fun and removed the challenge.

Peter Wilson recently told me that just after I had purchased the Mk I, Les Ryder was running a 2.4-litre engine test, with a long inlet manifold designed by George Buck. Using a small port head, the manifold was built up with Araldite, and then gas flowed by 'Windy' Windridge. Using 1.5-inch SU carburettors, it was developing about 133 bhp. During the test, Bill Wilkinson was heard to say, "Don't let that bugger, Martin, know what we are doing, or he will have one on his car tomorrow!" It was tongue-in-cheek of course, but what he didn't know was that I already had a version of it on my car, and it was working just as well as theirs!

A little later, in 1962, the JAMC had been invited to a sprint event, organised, I think, by the North Staffs Motor Club, at Curborough, near Lichfield. This would be my first competitive event with the Mk I, and I had to run in the modified saloon class because of the changes. Despite this, I came second to a very quick 3.4-litre, beating several others in the process.

By this point, I had decided that the van had to go, to recover at least some of the expense incurred by the purchase of the Jaguar.

I have never been able to avoid the pangs of remorse I get when parting with a vehicle, particularly one that I had owned for a long time, and had such fun

The Daimler Dart, later to be renamed the SP250.
(Courtesy John Starkey/Jaguar Daimler Heritage Trust)

with. It was with a sad heart that I placed an ad for the Ford 100E van, turned estate, turned rally car, in the *Coventry Evening Telegraph*. My asking price was £225, which I did not expect to get, but there was no harm in trying. Even after a repeat of the advertisement the following week, there were no takers. I was somewhat dismayed, until it was suggested that I put a for sale notice on the works notice board. It was free, and there was nothing to lose. Then, just three days later, I received a telephone call from a prospective punter, Stan, and arranged for a demonstration.

It turned out that Stan's father worked in the sawmill at Browns Lane, and had seen the notice in the works canteen. When he told Stan about it, Stan realised he knew the vehicle by sight and reputation, and thought it might be perfect as his son's first car. Stan, his son, and I, met in the car park of The White Lion in Brownshill Green, and they seemed impressed with the van's performance. They made an offer of £190 in cash. After a little haggling, we eventually shook on £210, which was more than I had hoped for, and the van and I sadly parted company. I kept in touch with Stan, and received reports from him every now and then. When he retired, we lost touch, and I often wonder what happened to the van after that.

Back in the experimental department, we had now started to build two cars that, upon their introduction to the public, would see Jaguar well and truly rise to fame and become an immortal part of automotive history.

Both of these cars would feature in the launch of the E-Type at the Geneva Motor Show on March 15, 1961. The first was the second FHC built, and would be road registered as 9600 HP. The second car was a roadster, and would also bear a famous registration number: 77 RW.

The history of these two press demonstration cars is already well documented, and I would suggest reading the appropriate section in the biography, *Norman Dewis of Jaguar: Developing the Legend*, where his dash across Europe in 77 RW is described in detail.

77 RW was desperately needed to support an already overwhelmed 9600 HP, such was the frenzy of interest in Jaguar's new 'supercar.' During the five years that I spent working as a volunteer at JDHT after my retirement, I drove 77 RW on several occasions. The car had been completely rebuilt after being found in a sad, sorry state years before, and is meticulously kept in its original specification for all to see.

By this point, Jaguar had taken over Daimler, and the factory at Sandy Lane was sorely needed for Jaguar's expansion plans. Gradually, the machine shop and the production engine assembly department, together with their sub departments, were moved out of Browns Lane to create space for the assembly lines to be extended. They would be followed by axle and front suspension build. This would mean major components would have to be delivered to Browns Lane, which eventually led to Browns Lane being labelled as a vehicle assembly plant.

The personal downside to these moves was that the little jobs that I was used to doing in the factory when things were quiet, such as chroming, hardening, painting etc, were now almost impossible to get done. Eventually, the Daimler factory would cease to build motor vehicles, and would become the home of Jaguar's production engineering. It is sad to think that these two great factories, with their wealth of history, have now gone forever, almost as though they had never existed.

Another advantage of Jaguar's move to Sandy Lane was the introduction of its own fibreglass department. Daimler had been quite forward-thinking in establishing this department, which they needed in order to build the body for the Dart (later to become the SP250, for legal reasons). To tool up for steel or aluminium would have been a prohibitive cost, especially at a time when Daimler was desperately short of money. What little they did have was being spent faster than it could be made, by Bernard Docker, a member of the Daimler board, and his wife Lady Docker, on lavish cars, among other things. So Daimler took a leaf out of Colin Chapman's book, and went down the fibreglass route. The SP250 body was beautifully made and very strong, but the engineering was woeful and the build quality poor. I remember Norman road testing one of the early production SPs, returning with a shake of his head. He had never witnessed scuttle shake like that before. One of the first modifications Jaguar made to the body, was to stiffen up the scuttle and therefore reduce the shake.

The addition of the fibreglass department meant that the E-Type roadster could have a beautifully made, coloured fibreglass hardtop, fitted to the bodyshell using the same attachment points as the normal hood.

I am not sure whether it was available to the ordinary customer in any other colour, but I do recall seeing at least one production E-Type with a white gel coat on its hardtop. Of course, some owners would later have theirs painted to the match the colour of the car, but this was not normal practice at Jaguar.

It was at around this time that I made friends with Shaun Baker. He was an ex-apprentice who stayed on as an assistant metallurgist, working in the Jaguar laboratory. There he did regular testing on materials, particularly bought out components, to confirm quality and conformity to specification. At the time, they were mostly working with fibreglass.

Shaun was the son of Colonel Baker, who in his retirement owned and ran Wolvey Garage, near the village of that name, quite close to the A5, near Hinckley. Shaun and I would spend a great deal of our time there, looking after and servicing

Colonel Baker's Wolvey Garage. (Author's collection)

our own cars, and helping the Colonel to run the garage.

Just down the road was the Wolvey Skid Centre, run by the Ogle family. The Ogle's had a fuel account with the Colonel, and it was through this that I met Peter Ogle, and was able to admire first-hand some of his futuristic designs for the motor industry. One of which, if my memory is correct, became the basis of Reliant's successful Scimitar.

I lost touch with Shaun and the Colonel after I left Jaguar for the first time in 1966. Sadly, Shaun was later killed in a road accident in South Africa.

Chapter 12

1961-1962

The Mk X, more racing, and the early lightweights

The E-Type was not the only vehicle under development. As I expect happens in most development departments, there were always several overlapping projects in their various stages of evolution. For us, one of these was the Mk X, Jaguar's first unitary-construction large saloon. Given the code name, 'Zenith,' it had a relatively short gestation period before its launch on October 11, 1961.

For the first time, a major dealer convention was held in the front car park of Browns Lane. Almost the entirety of the parking area was covered by a mass of interconnecting marquees for this grand event. This caused some discontent as this was normally the staff car park, and suddenly, parking spaces were hard to come by.

Before this, though, the Mk X would go through the most exhaustive test regime that Jaguar had ever performed. For the first time, road testing would be carried out on the roads of mainland Europe. One of the engineers who took part in the test programme (which is comprehensively described in Norman Dewis's previously mentioned book) was Barrie Woods.

The test route started in Bayonne, France, and went south-east, through the Pyrenees, to Oloron. There it turned north, via Pau, to Labouheyre on the N10. Then it returned south, back to Bayonne, a distance of 225 miles. Coincidentally, the route passed within a short distance of Monlaur Bernet, where my wife and I later resided when we moved to France. Two cars were involved in the test programme: 5437 RW and 5438 RW. Unfortunately, 5437 RW was lost when apprentice Fred Merrill had a major argument with a tree. Fred, however, came out almost unscathed fortunately.

Back in Coventry, the development programme on Zenith continued. Electrically, there were not many changes to what had now become a fairly standard specification. I suppose the major one was the change to a negative earth system, necessitated by the first-time use of an alternator, which replaced the dynamo. Now, we had the facility of much higher charge rates, particularly at low rpm. The increasing loads imposed by the addition of new systems made this particularly welcome: air conditioning, central locking, electrically-operated windows, and additional lighting were all taking their toll on the poor dynamo. The Zenith test programme warned us of ever-rising temperatures under the bonnet, and the effect that this would have on the delicate electrical components, especially the battery.

This problem would become much more severe with the introduction of the V12 power unit, which not only generated more heat, but also filled the engine bay more effectively, leaving less room for airflow. In early heat tests on the XJ12, the temperatures were so severe that the sealing pitch around the battery top melted, and the casing was distorted. I remember going to an engineering

Jaguar's new big saloon, the Mk X. *(Courtesy John Starkey/Jaguar Daimler Heritage Trust)*

Me in an XK150 FHC at Wellesbourne airfield. This particular XK belonged to a friend of mine, and we shared it for sprints. (Author's collection)

meeting where this problem was discussed. Several weird ideas were mooted. One suggestion was to run the air conditioning system through a box which contained the battery. Another was to provide the battery with its own cooling system in the shape of a cooling chamber and fan. After further development, this idea was eventually adopted for production on the XJ12.

During a lull in the conversation I said, "Why not just put the battery in the boot?" This was followed by a stunned silence, then everyone talking at once. "Oh, we cannot lose boot space," and "What about all of the extra weight, with battery cables the length of the cabin?" I said that this was not a new idea. We had done it before with the Mk II race cars, and surely the extra weight would help to offset the additional mass of the big V12 engine up front. As far as boot space was concerned, an engineering solution could be found. I was crying into the wilderness, and it would be some years before Jaguar's batteries would find a permanent home in the boot, the obvious place to put them; so much for progress!

For some time now, Jaguar had been considering the introduction of a new engine to support its aging six-cylinder unit, and they were looking at both V8 and V12 options. Jaguar had, almost by default, inherited a range of V8 engines. These were designed by Edward Turner for Daimler, and then produced in 2.5-litre form for the SP250, and 4.5-litre for the Majestic Major. Both of these engines were impressive units, particularly the larger capacity one, with a fair amount of 'grunt.' The downsides of the big V8 were its weight and its rough running at low revolutions, mainly due to poor casting and machining tolerances on the cylinder heads. Jaguar's own experimental V8s also suffered from rough running when compared with the smoothness of the six-cylinder unit. As most will know, eventually the decision was to try a totally new concept, which would end with the magnificent V12. Before that, however, we did run a Zenith with the large Daimler unit installed. We used an early Zenith, a rather dingy-looking black car, which, despite its bulk, would give a 3.8-litre E-Type a run for its money before eventually being overcome by its brick-like shape and high wind resistance. I recall driving this car on the M1 during testing, and seeing the startled look on the face of an Aston Martin driver (who was testing out of Newport Pagnell) as we passed him at about 130mph!

My main involvement on the Mk X was with the air conditioning, for which we

Jaguar Mk Xs outside the Browns Lane office, awaiting collection. (Courtesy Jaguar Daimler Heritage Trust)

considered units from several manufacturers, among them DeLaney Gallay and Martinair. The early compressors required a great deal of horsepower to drive them, and the vacuum-operated controls were notoriously unreliable. Large fan motor units, and the inherent bulk of these early systems, made access for service and repair very difficult. Suddenly, motor cars were getting very complicated.

The Mk X has never been a favourite of mine. There is no doubting its impressive performance and comfort, even in 4.2-litre form, but its sheer bulk and fat waistline did little for me.

It was during this period that another of Barrie's pranks went wrong, and caught him out in the end. Barrie's bench backed on to one used by Geoff Faulkener, who was a brilliant welder. On this particular occasion, Geoff was welding an exhaust manifold, which was clamped in the vice on his bench. Barrie had constructed a very large and lifelike spider from a lump of dumdum, with pipe cleaners for legs. Using a piece of string, he lowered his spider over the front of Geoff's bench, so that it passed in front of his welding mask. Geoff took one look at it, dropped everything, and jumped back, tripping over the welding pipes in the process. Crashing to the floor, he was knocked out when his head hit the welding bottles.

In the mayhem that followed, Geoff was rushed to the surgery. He returned a little later, fortunately suffering only a large bump to the head, but with a savage expression on his face. Meanwhile, Barrie had hastily retreated to the toilets, and would not return until Geoff had calmed down.

Weeks later, when everything had been forgotten, Barrie was in the main shop, lying on his back under the gearbox of a car on axle stands. He failed to notice that Geoff had recovered the spider and had been waiting for the opportunity to get his own back. Using a long piece of wire, he carefully pushed the spider along the floor until it came into Barrie's peripheral vision about 6 inches from his head. Startled, Barrie leapt up, smashing his head on the sump plug of the gearbox. Responding to his cries, we dragged him out by his feet, and saw blood pouring from a cut on his forehead. After his own trip to the surgery, and then to Coventry and Warwickshire Hospital for stitches, Barrie returned to his bench looking pale and very subdued. Although it may not seem like it, Geoff and Barrie were the best of friends.

There were many other memorable moments and stupid pranks, most of which would cause absolute havoc with today's health and safety regulations. One of my favourites involved the large coil spring from inside a brake booster. At the time, most mechanics used old army ammunition boxes for their tools, which had a shallow tray in the lid. The spring would be put under this tray, and would be compressed when the lid was shut and latched. The unsuspecting owner, when he next opened his tool box, would be confronted by a flying tray and an array of small tools. When I look back, I often wonder how we managed to get any work done at all!

Bill 'Nicko' Nicholson was still running the Alpine at this stage, and was loudly boasting that he could remove the engine in just ten minutes. One day, Barrie decided to give him the opportunity to prove it by pouring a cup full of engine oil underneath the car, near the bell housing. He then innocently stopped Bill in the workshop and said, "Oh, Bill, did you know that your car is leaking oil?" "No it isn't, my car does not leak oil," Bill replied. Barrie said, "Well there is a pool underneath it."

At this, Bill rushed outside and stared aghast at the evidence. He then said to Barrie "I will have to take it around the back of the shop and remove the engine. It will only take ten minutes." Knowing that the joke had gone far enough, Barrie told Bill what h'd done, to which Bill replied, "You can't fool me, I knew that it was a joke. That's Esso and I use Castrol R."

Yet another ongoing project at this time was the installation of the 'little' Daimler V8 into one of our Mk I test cars. A second prototype would see the 2.5-litre Daimler V8 put in a Mk II body. This would later become the Daimler 250 V8, the only production vehicle to marry a Daimler power unit with a Jaguar bodyshell.

In this project, there was very little work for the electrical department, other than minor changes to the wiring. The reduction in weight, increase in power, improved torque, and a manual gearbox made it a sprightly performer. It had better acceleration than its Jaguar sibling, as well as a higher top speed and improved handling. Only a very small number of this model were produced in manual transmission form, most of which were towards the end of its production. It was rumoured, at the time, that Sir William refused to offer the car in manual form at first because it might have affected sales of the Jaguar 2.4-litre. Personally, I have always doubted this, as the profits from the Daimler would certainly have been equal to, if not better than, that of its Jaguar sibling.

Despite Jaguar's exit from competition, there had been continued works support, as well as financial aid, for those owners who were capable of competing. Jaguar may have been supportive but they were never a benevolent society. Pressure was building, from a number of private sources, for a competition version of the E-Type, suitable for both national and international sports car racing. Some of these were long-term Jaguar competitors in saloon car racing, driving Mk Is and Mk IIs. Prominent among them were John Coombs, who's Guildford based Jaguar dealership had been competing with Jaguars from the days of the C-Type, and Tommy Sopwith's Equipe Endeavour, which had been very successful with the Mk II. What followed would be the start of the E-Type in serious competition. Before that, however, one very special car would be built.

A new face had appeared in engineering. Derek White was, I think, South African by birth, and was a brilliant engineer. In his spare time he built and raced a series of well-designed, Austin 7 based 750 specials. At Jaguar he would oversee several programmes, one of which was the low drag coupé registered CUT 7, made famous

CUT 7, one of the 12 lightweight E-Types, awaiting collection after service. (Author's collection)

by Dick Protheroe. For some time now, Malcolm Sayer had been formulating modifications to the shape of the E-Type, which incorporated his theories on how to reduce drag; they eventually came to fruition during the build of CUT 7.

I will not dwell too much on the build of this vehicle, as it has already been covered by others, but I would recommend Peter Wilson's book, *Cat out of the Bag: Jaguar–the Competition Department 1961-1966*.

My part in all this was minimal. The base car was already complete and very little electrical work was required in the conversion. The screen wiper system needed revision, to take into account the new windscreen shape and rake, and additional wiring was required for the axle cooling system, the revised scuttle, and the fuel injection pump. There was also a duplication of some of the important circuits (headlights, sidelights, etc), but apart from that the car was pretty standard in terms of the electrics.

However, this was not the limit of my contributions to the varying projects that went through the competition shop. I spent as much time as I could possibly justify in the comp shop, to the extent that if anyone needed me, that was the first place that people would look. Even Phil Weaver accepted me and was encouraging of my various forays into his department. In later life, after his retirement, he told me that he had appreciated my enthusiasm, and was well aware of my desire to participate in what was happening. The result was that I was given opportunities to get involved with many things which were not within my remit.

One of these unofficial projects was the build of the instrument panel on CUT 7. On this car, everything possible was being done to reduce the weight, and in my innocence I told Phil that I thought we could lose a little by throwing away the standard side fascias and centre dash panel, which were steel pressings covered with Ambla. I suggested substituting them with identical units made of thin gauge aluminium painted crackle black.

Phil thought for a moment and then said, "Well, you had better get on with it then." I needed no further urging, but I did have the presence of mind to discuss the subject with Bob Blake. Bob told me that the first thing to do was make plywood formers using the engineering drawings for the standard component, less the metal thickness. I could then use these to form the outer shape of the panels.

It was now that I discovered how infectious the thought of a Jaguar competition car was to the rest of the factory. I only had to mention that I was working on the new racer and doors automatically opened. I drew plywood from the sawmill, 16 SWG aluminium sheet from the metal stores, and a wicked-looking trepanning tool (to cut the holes for the speedometer, revolution counter, and small instruments) from the tool stores; I was in my element! Bill Cassidy, meanwhile, wanted to know why I was spending so much time in the competition shop and was suitably mollified when I told him.

The all-conquering Ferrari 250 GTO from 1962. (Courtesy John Starkey)

Having completed the manufacture, and carried out a trial fit in the car, all that remained was the finish. Bob helped me out again and told me to take the panels to the service paint shop. So off I went, panels in hand, to see Bill Norbury, the service manager. I explained what I needed and he sent me to see one of the paint finishers who would apply the crackle finish.

I watched, fascinated, as the process began. The panels were scrubbed and dried to remove any contamination, then they were heated in an oven before being sprayed with yellow etching primer. Another quick rub and clean was followed by the application of the special crackle paint, and then it was back into the oven. The heat shrank the paint to create the crackle, and I was overjoyed with the result. All that was left to do was to assemble all the instruments and switches, and fit it all to the car. When I stood back to admire my work, Bob Blake's voice came over my shoulder: "You made a good job of that." What an accolade!

The next car we built was the Coombs semi-lightweight E-Type. I say 'semi' to differentiate between this short run of steel competition cars, and the full lightweights, with their aluminium structure. This car would be registered 4 WPD, and would eventually transform into a full lightweight with the same registration number. Before that though, 4 WPD would be pitted against the strongest of oppositions from both Ferrari and Aston Martin, mainly driven by Graham Hill or Roy Salvadori. The E-Type usually had the legs of the Aston, but the Ferrari was a different matter, especially those 250 GTOs entered by Rob Walker and Maranello Concessionaires, and driven by the likes of Stirling Moss and Innes Ireland.

4 WPD was given the lightweight steel specification, introduced for CUT 7, when its original bodyshell was written off in Roy Salvadori's massive off-track excursion at the Goodwood Easter Monday meeting on April 23, 1962. The replacement 4 WPD was hastily prepared for the 25-lap GT race at the Silverstone International Trophy meeting on May 12, 1962. It gained a lightweight steel monocoque (which had been stored after the completion of CUT 7 and the cancellation of any further low drag models), as did several others, in particular the Tommy Atkins car, and the one prepared for Tommy Sopwith's Equipe Endeavour.

I remember one late-night saga that occurred, which involved a car for Tommy Sopwith, registered ECD 400. It had returned to the factory with a problem that required attention to either the axle or the gearbox, for which we needed a trolley jack with a vertical supporting post. For some reason, Bill Nicholson was acting as overseer while the work was being completed. As he lowered the ramp to put the car back on the floor, he was in the middle of a heated discussion with Bill Cassidy and had forgotten that the trolley jack and post were still under the car. I saw what happened next from a distance. The rear screws of the ramp failed to overcome the obstacle in their path, but the front screws kept going until gravity took over. The car shot off the front of the ramp straight into the lathe machine stood in front of it. This did the bonnet of the E-Type no good at all, and moved the lathe all of 3 inches across the floor.

It was absolute mayhem, with much criticism aimed at the unfortunate red-faced Bill, who for once was lost for words. Not only did we now have a severely bent motor car, it was embedded in the lathe, and still half on the ramp having fallen off the trolley jack in the process. The rear screws of the ramp had to be dismantled and wound back individually until they were in line with the front again. Meanwhile, the car was returned to the support of the trolley jack, so that its rear end weight was not imposed on the ramp. By lowering the back while

Me trying hard with the XK150 during a sprint at Wellesbourne airfield, Warwickshire. (Author's collection)

jacking up the front, we were able to manoeuvre the car back on to the ramp. We left the lathe where it was and the operator didn't even notice until it was pointed out to him several days later. All that was left to do was repair the car, which was required for an extensive test programme the next morning. The bonnet was well beyond immediate repair, and it was obvious that a replacement would be required, but where to find one at ten o'clock at night?

Most enlightened E-Type owners will know that you cannot just swap bonnets from car to car. It was never that simple. For instance, in production, all bonnets were about 20mm longer where they met the bulkhead of the monocoque. The assembly line operation was to pack and adjust the bonnet hinges until the front wheelarch profile was correct and fit as closely as possible against the lower edge of the monocoque. Only then would the excess be trimmed off with a pair of large hand shears, and the bonnet would finally sit in the area which was awaiting the seal: a rather crude but effective operation. Even if we were able to 'steal' a production bonnet, it would still be unpainted. The only option we had was to swap the bonnet, which would obviously require a great deal of fiddling. Eventually we got it to fit reasonably well, and the red car was ready for the test session the following day sporting a dark blue bonnet. I'm sure that raised some eyebrows and left Bill with some explaining to do.

It had been a while since Barrie Woods had last played a prank, and he decided that it was time for another. I had been joined in the electrical department, as previously mentioned, by Tony Boden. Tony was a great mate, but a little short on the grey matter, and easily lead into dubious areas. Just outside the electrical shop was the silent booth, which was occasionally used to remotely run an engine, for example, and exclude external noise interference. It had two large doors, similar to those on a shipping container. A steel girder, on which a chain hoist and sling was mounted, ran through a hole in the upper part of these doors.

Barrie decided that he would pretend to hang himself from this girder using a rope fastened to his belt behind his back, with a loose loop around his neck. An accomplice then persuaded the unfortunate Tony to open the doors, and he was confronted with a purple-faced Barrie, tongue protruding, 'hanging' from the girder. Tony passed out on the spot and, yet again, Barrie's prank ended with surgery. Tony never quite forgave Barrie for the cruel trick, and it would be some time before he even came around to speaking to him again. This sort of behaviour, of course, would be totally unacceptable in this day and age. It was not only downright dangerous, but also naive. However, things were different then; we worked hard and played hard.

In early 1962, we built one more special E-Type for that year's Le Mans race. Briggs Cunningham had, as usual, completed his entry for this event. His best bet for an overall win would come from a 4-litre Maserati sports racing prototype. However, he wanted to support this with an entry in the production GT category, and what better than an E-Type. Briggs had always been a faithful supporter of the marque, and had successfully competed with a roadster in the States. His Le Mans entry, however, would be an FHC, and to this end he purchased a completed production car from Jaguar with a remit for it to be prepared by the works for the race.

After a great deal of consideration, and some strong argument from Phil Weaver, engineering agreed that it would be cheaper and easier to build a new car from scratch rather than strip and rebuild a production vehicle. The production car was therefore returned to sales, and a brand new bodyshell was delivered to the competition shop for modification before paint; this would become 1337 VC. The meticulous build of the car is admirably covered in Peter Wilson's aforementioned book, and it incorporated all of the tried and tested systems that had been carried out on E2A, 4 WPD and CUT 7.

One of these was simply keeping the temperature down on the rear axle. We realised that heat soak from the inboard disc brakes, if not controlled, would be very detrimental to the axle seals. A reliable solution was needed because the critical corners of Le Mans, such as Mulsanne, Arnage, and Tertre Rouge, demanded constant heavy braking, and because of the race's duration, endurance was key. The solution came in the form of a U-shaped copper fin oil cooler, mounted beneath the floor in the airstream passing under the car. The axle back plate was modified to incorporate a flow and return system, and the flow pressure was supplied by an SU fuel pump. A capillary temperature gauge was fitted in the instrument panel and to keep the operation simple, the whole system was controlled by a switch on the panel. All the driver needed to do was monitor the temperature and turn on the pump when necessary. It may seem crude but it was very effective.

Another area that I paid particular attention to was the external lighting, which was critical. The last thing you'd want in a 24-hour race is dodgy electrics. The car was to be fitted with seven-inch Lucas Le Mans headlamps with 100/100 watt bulbs, exactly as the D-Type had, with bulb holders which could be easily removed from the rear of the outer shell for quick replacement.

We had already suffered switch failure because of the additional loads, and I therefore decided, where possible, that the important circuits would be controlled by relays, thus reducing the switch loadings. These included side, tail, and headlights, wiper motor, fuel pump, and axle pump, all fused and relay-controlled independently.

Added to this, the most critical circuits (lighting, ignition, and fuel supply) were provided with dual circuitry so that in the event of a system failure the connections could be swapped over. This might seem like overcomplication but long distance races like Le Mans have been lost many times by simple electrical failures, so it made sense to do all we could to avoid them.

The car ran like a dream in the race, finishing fourth overall and second in the GT category, with no electrical problems – job done! A great deal of time, effort, and thought went into the build of this car, and it was probably the most rewarding and successful projects that went through the competition shop. Roy Salvadori, who co-drove with Briggs Cunningham, said afterwards that it was one of the most comfortable and easy-to-drive race cars that he had ever competed in.

Briggs was so enamoured with the car that he decided to keep it, and after a strip down and rebuild, where practically nothing needed to be replaced, the car was shipped to America. It was campaigned in the States for a while and then retired to Cunningham's collection.

To this day, I do not quite know how I managed to get myself on the official support team for the TT at Goodwood on August 18, 1962. This 100-mile race for GT cars would pit the E-Type against its nemesis, the Ferrari 250 GTO, once more. Works support, in the shape of mechanics, Bob Penney, Peter Jones, and engine man Frank Rainbow, were assisted by myself and Peter Wilson, with Derek White in charge. We had long ago resigned ourselves to the fact that the E-Type, even in its lightweight, full factory specification, was not a real match for a well driven GTO. It was still too heavy, even though its suspension was more sophisticated. The GTO handled better, was more nimble, and more forgiving to the driver.

For this meeting, 4 WPD would be driven by Roy Salvadori because Graham Hill had opted to drive the Coombs GTO. This car was supported by a trio of GTOs, the fastest among them being driven by Innes Ireland, followed by John Surtees, and then Mike Parkes. The works Aston Martins were also seen as serious contenders, although as it turned out, the DB4 GT Zagato could not compete with either the E-Type or the Ferrari.

The Aston Martin team made their race even more interesting by inadvertently deciding to burn down their pits whilst refuelling. This could have had serious consequences for Jaguar, as we were only one spot away from them. The fire was under control quickly though, and a partially burnt and foam-spattered Aston Martin eventually rejoined the race.

Despite heroic efforts, the flying GTOs had their day, and Salvadori eventually finished fourth, which would have been fifth had Surtees not crashed out toward the end of the race. This would be the last race for 4 WPD in its semi-lightweight form, and it would be some time before the number would once more enter the fray, this time as a full aluminium bodied lightweight.

One hilarious event occurred in 1962, which livened up an otherwise boring day. It was mid-afternoon and I was working with the Lucas Switchgear programme on the Series 1 XJ6 and the Series 3 E-Type. I was in the middle of a very time-consuming destruction test programme, with a test rig consisting of two small servo motors, which continually turned the rocker switches on and off whilst various loads were applied to the switch contacts. The test was to be conducted to the point of failure on ten switches chosen at random from the samples supplied, and I was beginning to go cross-eyed from watching the two servos drone backwards and forwards. Suddenly, I heard a loud *whump*, and the building visibly shook. For an instant it went deathly quiet, then mayhem arrived. First it was the fire alarm, then the sound of running feet and much shouting.

I rushed outside to see the outer doors of one of the engine test cells, which were next to our department, were burst open, followed by a rush of flame. At this point, I was passed by Bill Cassidy, fire extinguisher in his hands, who rushed up to the doorway then staggered back, coated from head to toe in foam from the fire extinguisher that Phil Weaver was aiming at the fire from the opposite door. He had managed to also cover himself and everyone else around him. The extinguishers had little effect, and even the works fire brigade, who arrived very quickly, were struggling until the Coventry Fire Brigade arrived. The problem was that the fire was being fed by an open fuel tap in the test cell and until that was isolated, it continued to burn.

Jaguar had been consumed by the fear of fire since 1957 and controls on inflammable materials were very strict. However, the authorities had not been prepared for Bill Duff.

Bill was a taciturn Scot who, along with Ron Beaty, made up an engine test team. It was their cell that had gone up in flames. He was also our union representative and always carried with him a small attaché case for his 'union papers.' What was unknown to the powers that be was that Bill had a little fiddle going, and it wasn't union papers in his case, it was his petrol can, which probably held about a gallon of fuel. This enabled Bill to run very economical vehicles topped up by 'free' petrol.

The controls for the test cell were housed in a large console, which contained various switches, instruments, and controls for the engine being tested. Buried behind this console, in the main fuel line to the engine, was a tap. This is where Bill stood his can to be topped up.

If my memory serves me correctly, Bill and Ron were running an endurance test on load, which meant that the Heenan & Froude dynamometer connected to the engine would be imposing a load equivalent to that of the car's transmission whilst the engine was run at various rpm. During these tests, the thin steel exhaust manifolds would become white hot.

Now, either Bill had been called away on union business, or he had simply forgotten about his can. The resulting overflow of fuel gathered in the ventilation ducts that ran beneath the engine, and eventually caused combustion when the vapour came into contact with the hot exhaust. What made the situation worse was that fuel had also flowed into the main storm drain that ran under the field in Browns Lane, behind the experimental department, and terminated in Pickford Brook! As the fire progressed, it blew off all of the manhole covers right down to the brook. It was more by luck than judgement that no one was seriously injured. I do not think that Bill's fiddle ever became public knowledge, but it was noticeable that he stopped using his attaché case after the fire.

Chapter 13

1963-1965

The lightweights, the XJ range, and a big decision

The end of 1962 and early 1963 would see Jaguar's last attempt to make the E-Type more successful in competition, especially against its most serious rival, the Ferrari 250 GTO. The cars produced would still be privately owned and paid for, although with a great deal of factory support in most cases. There was still no indication that Jaguar was about to officially go racing again.

This short run of vehicles, for very particular customers, would be the true lightweight E-Type. The first to become a full lightweight, and about to start its third life, was 4 WPD, the Coombs car, which had been campaigned with varying success by Graham Hill. First it was a modified production car, then a semi-lightweight (after the original body was written off at Goodwood), and soon to be a full lightweight.

The aluminium monocoques produced by Abbey Panels needed a great deal of rectification before they could be built up, particularly around the tail lamp and rear number plate apertures, where the aluminium was liable to developing splits during the pressing process. This is where Bob Blake and his welding expertise stepped in.

Most of the main components had already been proved, but a major change came with the introduction of an all-new alloy block engine, with 35/40 head, dry sump lubrication, Lucas mechanical fuel injection, and eventually a five-speed ZF gearbox. This would become the standard for future lightweights.

The reincarnation of 4 WPD was completed in early 1963, and, driven by Graham Hill, took the chequered flag on its first outing in March of that year. It is fair to say that this win was not that surprising given that the main competition was Roy Salvadori in the Atkins Cooper Monaco. Nevertheless, it was a positive start.

I now had an official engineering specification for the lightweight programme, which enabled me to order certain bought out components. However, most items, apart from those of a special nature, were drawn from production. This usually meant that I went round to the departments concerned and filled my box with what was needed. This was much easier than the usual signing off paperwork from one department to another, and anyway, the quantities concerned were not significant. To this day, I have doubts that the lightweights were ever properly costed, and I believe that Jaguar must have made a significant loss on the whole programme. It is amazing that today, the cars that we built and sold for a few thousand pounds in the early 1960s will now change hands for millions, such is the price of exclusivity.

Cars two and three were destined for the USA, one to the ever-faithful Briggs Cunningham, and the second to Kjell Qvale, the Jaguar West Coast distributor. Both of these raced at 12 Hours of Sebring in March 1963, with the Qvale car coming home seventh overall and the Cunningham car eighth. In the over three-litre class, however, they came first and second. Incidentally, the Qvale car became a true barn find when it was discovered in the garage of an American house in 1999 having spent thirty years in hibernation!

The electrical specifications of the lightweights were a little different to the standard production E-Types, the only major variation being the wiring for the Lucas 'bomb' fuel injection pump, which required a load-limiting relay, and an axle oil

Two lightweight E-Types at Mallory Park. On the left is 86 PJ, the car driven by Roy Salvadori, and later Roger Mac. On the right is the John Coombs car, 4 WPD, raced by drivers such as Graham Hill and Jackie Stewart. (Author's collection)

cooler pump. This changed, however, when we came to prepare the Cunningham lightweights for Le Mans.

Cars four and five were built for the Tommy Atkins stable, and Peter Lindner, who became European Saloon Car Champion in a Mk II, and was a Jaguar distributor in Germany. Car six was built for Peter Lumsden, only to be written off in its first race at Nürburgring. The car was returned to Browns Lane and fitted with a brand new monocoque in time for the 1963 Le Mans.

For that event, Briggs Cunningham entered a team of three lightweights, which was made up of cars seven and eight, and the original car two. They were road registered as 5114 WK, 5115 WK, and 5116 WK. Despite heroic efforts from the pit crews during the race, only one car finished. Roy Salvadori got caught up in a monumental accident on the Mulsanne Straight, effectively destroying 5116 WK, although fortunately causing little damage to himself.

For the first time, these cars were fitted with Jaguar's new synchromesh gearbox, which was intended for production with the 4.2-litre engine in the Mk X and Series 2 E-Type. 5114 WK succumbed to gearbox failure early in the race, and 5115 WK was hampered with similar problems until Sunday morning, when brake failure on the Mulsanne Straight resulted in a complete modification to the front of the car.

It has been recounted many times how Bob Grossman (the driver of 5115 WK) struggled to get the car back to the pits, and how the bonnet from 5114 WK would be used in a true cut-and-shut job to finally get the vehicle back in the race. The car eventually finished ninth overall, which was all that Briggs Cunningham had to show for a very expensive weekend.

Car number nine was built for ex-D-Type racer Peter Sutcliffe, closely followed by car number ten for the Australian driver, Bob Jane, who won the Australian Touring Car Championship in a Mk II.

Car number 11 was special in that it was not intended for racing but was bought by its owner, Dick Wilkins, for use as a road car. A few creature comforts were added, such as the standard production E-Type heating and ventilating system, a full set of carpets and trim, and even wind up windows. Undoubtedly, at the time, this would have been the quickest E-Type used on public roads anywhere in the world. The official run of full lightweight specification cars would finish early in 1964, with the build of car number 12 for UK hill climb expert, Phil Scragg. Thus ended a very busy period for the competition shop, although we had not seen the last of some of the cars as Jaguar would continue to support private entrants.

To me, building the lightweights was a little like being back on the assembly line, as each car was virtually identical, especially in terms of the electrics. However, these cars were each assembled with much more loving care and attention than a standard production car.

Meanwhile, the proximity to motor sport had left me with the desire to get involved but motor racing has never been a cheap hobby, and with my paltry funds participation seemed unlikely. However, I had made friends with another member of the Nuneaton Motor Club, who ran a small garage about a mile from where I lived. I had, for some time, helped in the workshop there to increase my bank balance and enable cheap repairs to my own vehicles.

One day, I was surprised to find a very sad-looking Austin Mini in the rear of the workshop. It looked as though it had been driven into a brick wall, as the front of the car was a mangled mess. I found out that it was an insurance write off, and my friend Bill, the garage owner, had bought it with the intention of rebuilding it for his daughter. As it happened, she had already obtained her own wheels, so the mini was pushed into the back of the workshop, where it sat forlornly for some time. The seed had been sown, though, and it was quickly germinating. After a short conversation with Bill and the exchange of £75, I became the owner of KEY 549, but what to do with it? My first thought was to prepare the car for rallying as circuit racing still did not seem financially viable. Even for rallying I had limited funds, so with a heavy heart I decided to sell the Jaguar.

Another member of the Nuneaton Motoring Club had been pestering me about the Jaguar for some time, and jumped at the opportunity when it arose. I cannot remember how much he gave me, but it was substantially more than what I had paid for it. The proceeds enabled me to purchase yet another everyday car, to use whilst I rebuilt the Mini. My first foreign car was an early 1100cc Volkswagen Beetle for which I paid £125.

The Beetle served me well for the short time I had it. Its two memorable occasions were a class win in the JAMC production car trial of 1964, and catching fire in the middle of Bodmin Moor, which could have ended with the demise of both the car and myself.

As usual, it was all my own fault; I have never been able to leave motor cars alone. The oft quoted statement 'if it ain't broke, don't fix it' never seemed to apply to me. I had converted the car to a 12-volt early in my ownership, and the new 12-volt battery lived under the rear seat cushion, exactly where the 6-volt had been. However, the 12-volt was slightly taller.

The VW Beetle in front of the GEC block at the factory. (Author's collection)

A friend and I had been down to the South West coast for the weekend, and were travelling back to Coventry over the moor, when I sensed a hesitation in the normal beat of the flat-four engine. I was not too concerned as it immediately picked up again. I glanced in the mirror and thought it was getting foggy – but fog does not smell like that. Suddenly, the engine died, and when I looked around again, the fog had taken on a red glow: fire!

"The damned car's on fire!" I shouted.

I rolled to a halt and we quickly exited the vehicle. The rear seat cushion was now well and truly alight, and omitting a noxious, black smoke. Access to the rear compartment was limited by the two-door configuration, but I managed to lift the drivers seat and release the rear cushion from its retainers. I dragged it out, still burning, on to the road side, where, fanned by the additional oxygen and wind, it continued to blaze merrily. It had also transferred its pyrotechnics to a small gorse bush. Now I had a potential moor fire on my hands as well!

Fortunately, my conversion also included a CTC fire extinguisher, and so I was able to prevent the destruction of Bodmin Moor, and rescue a very sad-looking Volkswagen rear seat cushion. On inspection, the cause of the fire became more obvious. Due to the new battery being taller than the previous one, the seat springs had rubbed across the connections, eventually breaking through the insulation and causing them to short out. The heat generated then set fire to the foam in the seat cushion. It was a very salutary lesson. As the rest of the car was undamaged, we were able to continue our journey, minus a rear seat.

I had already started work on the Mini conversion, which had resulted in a lot of late nights. The bodywork was quite straightforward: two new wings and a new front valance panel (both reproduction), repairs to the outer scuttle panel, plus a fibreglass bonnet. To avoid the embarrassment of the bonnet flying off, I hinged it at the front with detachable hinge pins, which meant the whole bonnet could be removed in seconds. Some people went even further and hinged the whole front of the car, which gave superb access, but I decided against that.

My Mini, race ready. (Author's collection)

With the repairs complete and the body now partially de-seamed, it was time for some paint. I was quite impressed with the finish I achieved given my limited experience and equipment. With a primrose yellow shell and black roof, the Mini was now looking quite smart.

My first entry into circuit racing was in 1963, and in my mind I was on my way to the World Championship. However, the first couple of events soon cured me of any delusions of grandeur. I discovered that there were many others with similar intentions, some of whom were much better drivers with much better cars. It was great fun though, and during that year entries at Mallory Park, Castle Combe, Crystal Palace, Brands Hatch, Silverstone, Thruxton, and Oulton Park, together with hill climbs at Prescott, Rest and Be Thankful, and the odd sprint, enabled me to collect enough signatures to gain a National licence.

I must say that the Mini performed exceptionally well and was ultra reliable. I was, of course, always conscious of the need to keep the car in one piece in order to get home afterwards!

The only mechanical failure occurred during the race at Oulton Park, when I found that a rocker arm had broken. Had it occurred in practice and not during the race, I would probably have been able to finish. I managed to replace the faulty item and return home unscathed.

My most interesting race was at Mallory Park for, I think, a Nottingham Car

The Mk II production line, 1964. Note the exhortation hanging from the roof! (Courtesy Jaguar Daimler Heritage Trust)

Club event. I had qualified for a third row start, and I knew that I had very little chance of a reasonable placing. In front of me were several 'hot' car/driver combinations, in particular Harry Radcliffe in the Buick V8-powered Vita Foam Mini, and Tony Pond in the Climax-engined Tasman Anglia. As the flag fell everyone started to move and then seemed to falter. To avoid shunting the car in front, I pulled to the inside, took to the grass and kept my foot down. Suddenly, I found myself with a clear track ahead. Glancing in the mirror I could see mayhem on the start line, and I had the mesmerising view of a Mini's complete chrome side trim suspended in mid air, along with a cloud of blue smoke.

I was in the lead of a motor race! I hurried through Gerard's and on to Stebbe Straight. Halfway down the straight, another glance in the mirror confirmed that I was not alone. There was an Anglia and a Mini, side by side, catching up very quickly. I managed to hold them off through the esses by filling as much of the track as possible, but then we went into the hairpin at Shaw's three abreast, with me in the middle. They say that three cars cannot get round that corner, but we managed, and around Devil's Elbow they both powered away from me.

I eventually finished fifth, which was not bad considering. On my return to the paddock, I was amazed to find tyre scuff marks down both sides of the car, which shows how close together we were. Harry actually came over and apologised, telling me that the marks would probably polish out. They never did but secretly I was quite proud of them!

It was now 1964. The lightweight programme was slowly coming to an end and it was decided that works support would be concentrated on two particular vehicles: 4 WPD, the Coombs lightweight for UK GT events, and Peter Lindner's 4868 WK, which he would campaign in world championship GT races. Both cars would be redeveloped in an attempt to make them more competitive. In the case of 4 WPD, this led to large section wheels, modified wheelarches to suit, thicker roll bars, a reduction in ground clearance, and more power, all efforts to increase cornering speed. On straights, the E-Type had proved to be as quick, if not quicker, than the GTO, but it still lacked the ability to compete favourably against the more nimble Ferrari, which had a distinct advantage on corners.

Peter Lindner's car, however, would be different, and would be redeveloped substantially, along the lines of CUT 7. Lindner had an entry for Le Mans that year and so the car was returned to Browns Lane for the modifications to be carried out in time for test day in the middle of April.

There had been some changes in personnel prior to this period. Eric Bailey had moved into the GEC block to act as Burt Tattersall's assistant, which meant we now had two people in charge who knew little about vehicle electrics. Although, to be fair to Eric, he was far more informed than his boss!

Eric Wright, the former superintendent of the experimental department, had also moved on, and was now working for Rover at Solihull. He had been replaced by Phil Weaver, who still held responsibility for the competition shop, and Ted Brooks had at last been rewarded for his service to the company by being made foreman.

In some ways, for me, this latter change created some difficulty, as I didn't feel that Ted was as approachable as Phil. On reflection, I suspect that Ted did not adjust well to his appointment. He had been suffering for some time with stomach ulcers which suddenly became much worse. His diet during the day seemed to consist of milk and bananas, and he became more irritable. I found it difficult to spend any time in his department unless on a specific project, a situation that I was not used to.

However, this didn't cause me too much trouble when working on the Lindner car, as there was a substantial amount of electrical work to be carried out in preparation for the 24-hour race. This meant that when 4868 WK was returned to the competition shop, it was stripped out, and its engine was returned to the experimental engine department for redevelopment; I had no need to go to the competition shop. The first major problem arose when it was discovered that the formers, which Abbey Panels had used to produced the fastback roof section for CUT 7, had been disposed of. This was a major setback as it meant that all of Malcolm Sayer's original drawings and details would have to be reproduced to make a new set of formers before Abbey Panels could then make another roof section.

Meanwhile, I drew up a specification, very similar to the Cunningham lightweights, with the duplication of circuits where necessary, the addition of two auxiliary driving lights in the bonnet, a pair of additional tail lights, and illumination for the competition numbers. The car was ready to be despatched for the test weekend, where Peter Lindner would have his friend and partner, Peter Nöcker, as co-driver. Neither driver had competed at Le Mans before, but they produced some credible lap times during the first tests, getting down to 4 minutes and 7.3 seconds on the second day. However, the car was due to compete in one race before the main event on the Sarthe circuit. This would be the Nürburgring 1000 km on May 31. With this in mind, the car was returned to Browns Lane for further preparation.

In an attempt to clean up the front of the bonnet, and therefore reduce drag, I removed the side/flasher units and modified them to fit inside the bonnet structure, with a flat amber Perspex lens in front of each, flush with the outer skin. Further work to reduce drag was carried out as per Malcolm Sayer's instructions, and finally the car was ready and despatched to Germany for the race. The car performed well in practice for the event, but disaster hit halfway through the race, when the gearbox broke, eventually failing completely. This did not bode well for Le Mans. The car was returned to the competition shop once again for final preparations.

The race itself proved an anticlimactic and expensive weekend for Lindner and Nöcker. After four hours, Nöcker brought in the car for a scheduled stop and reported a water temperature of 100 degrees Celsius. Upon checking, the engine was found to be devoid of water. This situation continued through the night: refilling after the required laps, until the head gasket was eventually changed. Before long the same problems occured, and at 7.30am on the Sunday, after consuming another two gallons of water and now requiring oil, the car was withdrawn. It was a sad end to Jaguar's illustrious career at Le Mans, and it would be many years before the company once again found glory at the Sarthe.

The car was once more returned to Browns Lane. When the engine was removed, the reason for the failure was obvious. The aluminium block had a large crack in its rear face, undoubtedly due to flexing between its mounting points. The fitting of the heavy ZF gearbox, with its long first motion shaft, had required a thick aluminium spacer to be placed between the gearbox and the bell housing, thus increasing the unsupported length of the whole unit. The aluminium block

LWK 707, an XK120 FHC, in the Jaguar Heritage Museum. This car broke several records at Montlhéry by averaging a speed of 100mph for seven nights and seven days. (Courtesy John Starkey/Jaguar Daimler Heritage Trust)

was simply unable to take this load and eventually 'broke its back.' Had the heavier iron block been retained, this problem may have been avoided altogether.

This is the way that 4868 WK would be prepared for its next race at Zolder – with an iron block and a Jaguar four-speed all synchro gearbox in place of the ZF – only to fail again in the hands of Peter Nöcker. Failure, yet again, was the result of an entry in the Goodwood TT, where, having been invited to try the car in practice, Peter Sutcliffe only succeeded in stuffing it into a bank, thus declaring it to a non-starter. On its ignominious return to Browns Lane, the car would prepared for one more race. What we didn't know was that this would be the last time that a works built car would leave the competition shop, and it would have tragic consequences.

The Paris 1000 Kilometres in 1964 was held on the banked circuit at Montlhéry, the scene of a Jaguar triumph in 1952 when LWK 707, an XK120 FHC, ran for seven days and seven nights at an average speed of over 100mph. There were to be no such accolades for 4868 WK, driven again by Peter Lindner. The event was held in atrociously wet conditions and the car was involved in a high-speed accident with another vehicle, which, tragically, killed Peter and four others. The vehicle was completely destroyed, and photographs of the remains clearly show how comprehensive that destruction was. Thus ended an era of Jaguar-prepared cars in competition.

After being impounded by the French authorities, 4868 WK remained in a lock-up garage at the Montlhéry circuit for seventeen years before being released. In 1982, when the vehicle left the Lynx factory, very little of the original car remained, having been built up from one of the spare monocoques that Lynx had acquired when they bought the remaining spares from Jaguar. It is now seen as a reminder of what nearly was.

The days of the front-engine GT car were over, and from now on, the cart would be before the horse in this and most other aspects of motor racing. Jaguar was not behind in this respect. For several years, Malcolm Sayer had mused on the concept of a mid or rear-engine sports car, and had produced models and sketches that demonstrated his thoughts. By the middle of 1964 these thoughts had become a specification, and by August of the same year the XJ13 was under construction.

I first heard about the XJ13 project through Derek White, with whom I had established a good rapport, possibly because we were both members of the 750 Motor Club, and he had been running 750 specials for some time. He had approached me in June of 1964 and told me what was to come. Electrically there would be little difference between the XJ13 and what had gone before it. We were not yet into microchips and ECUs, but they were fast approaching. We would, for the first time, be using the Lucas Opus electronic ignition system, which basically did away with the old method (producing the spark via a coil and a set of points), whilst retaining the mechanical distribution of it. The Lucas 'bomb' fuel injection pump would also be used. Both of these items were of Bosch ancestry and produced under licence by Lucas. Everything else would have to be hand built, as this was essentially a one-off project with the hope of more to follow. We mere mortals on the shop floor were not to know of the plans which would eventually make the XJ13 a beautiful white elephant, a whisper of what might have been.

I approached Burt Tattersall to talk about a specification for this important project. To say that he was annoyed would be a massive understatement. He was livid. He knew nothing about this project, and even as electrical development director, he had not been informed. I was to do nothing until instructed by him.

Suitably chastened, I returned to the electrical department and went back to testing systems for the XJ6. Some days later, Eric Bailey called and asked me to go to Burt's office. When I arrived, Eric was asked to leave. "Here we go," I thought, "how do I react to this?" Burt, who was much calmer now, proceeded to tell me that he had been to a meeting with Mr Haynes, where he had been informed of the XJ13 project. He said that as I was the only one in the electrical department with experience of the work required on competition cars, I would be released from other projects unless they became critical, and that I should liaise directly with Derek White (and later Mike Kimberley) in all things that related to the specification of the vehicle.

Burt had returned to his normal, ebullient self, and almost condescendingly wanted to know if I needed help. Like hell I did. I'd never needed it in the past

The XJ13, of which the factory built this sole example. (Courtesy John Starkey/Jaguar Daimler Heritage Trust)

and I was not about to start now. I would, however, keep him informed of progress with regular updates. Thus I started on project XJ13.

As usual for me, I would get involved in more things that were not in my remit, the first of which would happen before I got hands on with the XJ13; it would be late in 1964 before the assembly of the panels (made by Abbey Panels from the formers produced in the competition shop by Bob Blake, Geoff Joyce, and Roger Shelbourne, another Jaguar apprentice) would take place.

While I waited though, I was not idle. The floor pan of the car was being assembled beside Bob Blake's bench in the competition shop, constructed from the floor upwards. I was intrigued by the attention to detail on the main fuel supply system, and in particular the large square tube that crossed the cabin floor and linked the two sill mounted tanks. Bob explained the principle to me. The tube contained the centre fuel tank and linked the two sill mounted bag tanks. This allowed all the tanks to be filled from one point, and for fuel to be used from the centre, thus maintaining the weight distribution. A hinged flap valve at each end of the tube would close under cornering forces, preventing the sudden rush of fuel from one side to the other, and therefore preventing the car's balance from being upset. Simple but ingenious.

Initially, there was some pressure on the build programme, and we anticipated that the first car would run in the 1965 24-hour race very much as a 'feeler.' Then, if that showed promise, a team of at least three cars would be constructed for 1966. Derek White had been appointed project engineer, and both he and Malcolm Sayer spent a great deal of time on the shop floor.

Quite unexpectedly, with the monocoque incomplete, work came to a grinding halt. Derek's frustration was clear, as a debate developed over the design of the front suspension. Derek had intended it to be lightweight and fully adjustable, in line with accepted racing practice. However, Bill Haynes insisted that the production E-Type system should be used, with forged upper and lower wishbones linking a forged upright with standard ball joints. Eventually, being chief engineer, Haynes got his way, and Derek's tubular wishbones, with infinitely adjustable rose joints and fabricated uprights, were consigned to the bin.

I had yet to start work on the vehicle, and at this stage had only just set out the electrical specification. The instrument panel layout was similar, in most respects, to previous practices with the D-Type. The large speedometer and revolution counter were housed in an angled panel to the driver's left, with the smaller instruments scattered around. The switchgear would be of aircraft quality, manufactured by Arrow. The only additional circuits would be for the Opus system and the Lucas 'bomb' fuel pump.

With the construction of the instrument panel, I tried to design it so that there was as little lighting reflected in the sharply raked windscreen as possible, as this could be very disturbing for the driver in a high-speed race at night. In practice, this never worked, and the best advice that could be given to the driver was to turn the panel illumination off and only use it as a check when it was safe to do so.

Derek White never recovered his interest in the project following the overruling of his front suspension design, and when offered the post of chief designer at the Cooper Car Company he moved on. Thus, around the middle of the year, we were suddenly without a project design engineer. Bill Haynes solved this problem by appointing Mike Kimberley to the post. Mike was an ex-Jaguar apprentice, very talented, and a good organiser, who at that time was working in engineering as an engine design draughtsman. On leaving Jaguar, he would go on to become, among other things, managing director of Lotus. Unfortunately, XJ13 would never recover from the lost time and mismanagement. It was too late and already out of date by completion. In a way, and with hindsight, it was probably for the best. Racing car design was moving very quickly, and to build a successful team of cars would have required much more finance and dedication than Jaguar's management were able, or even willing, to give the project.

Despite the car now being a lost cause, everyone soldiered on, but the momentum and interest had gone. It also became obvious that the word had spread to the rest of engineering, and I had distinct problems in obtaining clearance from Burt Tattersall for some of the bought out items. My time spent in the competition shop was also being questioned. Now, more pressure was being put on the design of systems for project XJ4, Jaguar's all-new large saloon, which would eventually become the XJ6 and XJ12 (all very confusing).

Testing for some of the systems on the XJ4 was now under way. On this model, there were several electrical firsts for Jaguar, one of which was the introduction of multi-plug connectors for the main electrical harnesses through the bulkhead. For some time, engineering had been concerned about the number of piercings that were required in the bulkhead to carry cables, tubes, etc. They posed major problems in sealing and noise transmission. After a lot of thought, a system was devised whereby double-sided male plugs, which passed through the bulkhead, were permanently sealed in place, and carried a series of round multi-pins, which mated to female sockets on each end of the respective harness. In some respects this was a big step forward; for instance, if you needed to replace a forward engine bay harness, this could be done by simply pulling out the corresponding plug with no need to disturb anything inside the vehicle.

My reservations about the system were how it would perform in service, particularly with aging. I had fears that the pins could corrode, and that hardening and deterioration of the rubber moulding encasing them could lead to sealing and contact failures. Lucas had done its own testing and had every confidence in the system, but I had been down this road before. We thus carried out our own series of tests, that imposed excessive current loads under raised temperature conditions, and exceeded those to be found under the bonnet, whilst subjecting the assembly to water and oil. The performance turned out to be surprisingly good, and I do not recall any failure during these tests, which were designed to simulate 50,000 miles.

Another system that we tried on the XJ4 was a single wiper blade. The advantages of this were that it could be geared as a direct drive to the motor via a crank (thus avoiding the complication of wheel boxes), a flexible drive, and a reduced load on the motor. These advantages, however, were outweighed by the doubts that a single large wiper blade would be capable of clearing enough of the windscreen under extreme conditions, and the problem of locating the wiper motor in the already signed-off bulkhead. Jaguar did eventually go back to this system on the XJ40 and the X300/308. Although work was progressing on the XJ4, the car had, for Jaguar, a long gestation period, and it was another three years before the Series 1 saw the light of day.

After two seasons of campaigning the Mini in club racing, I decided it was time for a change. I accepted a substantial price from a man in Devon, and was sorry to see it go. Although I had never been really successful with the car in terms of race wins, I did usually manage to finish; it had rarely let me down and I had a lot of fun. At one time, a friend of mine had a 105E Anglia, and I had been impressed with both the performance and handling of this 997cc model; so much so that I decided to race one.

Racing in 1964 at Mallory Park. (Author's collection)

I managed to find a basic car, with good bodywork and a scrap engine – just what I needed as the engine would be replaced anyway. Over the next few months, in my rented lock-up garage, I stripped the car to a bare shell, modified the front valance for better brake cooling, and fitted the attachment points for the Watt's linkage and trailing arms on the rear suspension. I then welded and sealed the rear bulkhead, and bolted a small eight-gallon aluminium fuel tank to the boot floor. I painted the finished car in metallic silver blue, which looked very impressive.

The scrap 997cc engine and gearbox were discarded, and replaced with a 1600cc crossflow unit fitted with a Cosworth race cam, enlarged inlet valves, and a polished and gas flowed head. A pair of Weber 40 DCOE carburettors fed across to a four-branch tubular exhaust manifold. The front suspension retained the original MacPherson struts but with lowered springs and early Capri disc brakes, plus an additional anti-roll bar. The live rear axle was tied to the bodyshell by retaining the existing leaf springs, although with several leaves removed, located by a pair of rose jointed trailing arms to prevent axle wind up, and a Watt's linkage off the rear axle casing reduced lateral movement.

All of the chrome trim was removed, along with door casings, carpets, etc, and finally the side glass and rear window were replaced with Perspex. Wider wheels were not readily available for the Anglia, at least not at a price that suited my budget, so I decided to use standard Ford 5.5J x 13 at the front and make up my own seven-inch section wheels on the rear.

At this stage, the expertise of the experimental department came in handy. The machine shop was going through a particularly slack period, so Bill Cassidy gave me permission, so long as he was not involved, to use the lathes. The largest of the lathes was used to remove the rims of four Ford 13-inch wheels. Two had the rims removed from the outer side, and two from the inner side. These were then paired up to make seven-inch section wheels.

The next step was to weld

Above: Me trialling with a Ford Anglia. (Author's collection)

them together with the help of Harry Hawkins. Using a jig, Harry arc-welded the rims together, then, to be on the safe side, he welded four stiffening ribs on the outer side of the wheel, from the centre to the rim. The machining and welding were so accurate that very little metal had to be removed when the wheels were dynamically balanced. I now had a pair of seven-inch wide wheels for the cost of a few packets of cigarettes and just a bit of swearing from Harry.

The Anglia and I then went racing for a couple of seasons. This time, however, it would not be driven on the road, and so I acquired a very smart Triumph TR2 with a tow bar and a trailer as my newest toy, which allowed me to transport the Anglia to and from races.

After the first season of moderate success, I decided that more power was needed, and concluded that the most effective (and cheapest) way would be to supercharge the engine. I managed to find, for a modest sum, a second-hand Wade twin-rotor blower, which was strapped on top of the Ford power unit, and driven by twin belts from the nose of the crank, fed to a fabricated inlet manifold via a single two-inch SU carburettor. The height of the system necessitated a large cut-out in the bonnet, covered by an air scoop to assist in cooling the blower.

Despite generating huge lumps of power when the blower came in, the system was never really successful. It required more time in development than I was prepared for, and was stretching my finances to the limit already, so, with a heavy heart, the Anglia, its trailer, and I parted company in a pub car park just off the North Circular, and I returned to Coventry with the TR2 and a pocket full of cash for my next project. This, however, would take a few years to reach fruition as shortly after selling the Ford I met my wife to be, but that's another story.

With the end of the lightweight E-Type programme, and the completion of the XJ13, the competition department would not be solely focussed on racing projects for much longer, and it soon became obvious that it would cease to exist altogether. Although there was no indication from the management of what would happen, the worldly-wise among us could see that eventually things would have to change. We foresaw that the experimental engineering and design department as we knew it would disappear for ever, to be replaced, some time in the future, by an all-encompassing design and engineering facility which would be located in Whitley. That wouldn't happen until late 1965, but I was becoming disenchanted with my work life. Employees were beginning to get twitchy and unhappy about their future, and with very just cause; looming on the horizon were amalgamations and agreements that would eventually lead to the near extinction of Jaguar as a brand name, and the rise of British Leyland. We were soon to witness the fall of the British motor industry!

I now had no reason to be in the competition department (where little of interest was happening anyway), and the everyday grind of trying to find something interesting to do was beginning to bore me. At that point, I began to seriously look at what alternatives there were away from Jaguar. Eventually, in September 1965, I took the plunge and left for pastures new. This was not an easy decision; I was 32 years old and all of my working life to date had been spent with either Jaguar Cars or Her Majesty's Forces. It was definitely unknown territory.

Three weeks later I became the UK installations and service manager for an American company, selling and installing Frigidaire coin-operated launderettes; how the mighty have fallen!

It was not as bad as it sounds. This new life provided me with a much broader understanding of what went on outside the warm cocoon of my previous existence. I still had my contacts at Jaguar and the Jaguar Drivers Club, I had another two years of club racing to look forward to, and last but not least, I met the love of my life and married her.

However, this was not the end of the road for me and Jaguar. Fate had a few more twists and turns in store.

Chapter 14

1965-1972

New jobs, Forward Engineering, and coming full circle

While this chapter has little to do with Jaguar, this period was a major part of my life, and gave me a clearer understanding of what goes on in the real world. Until now I had walked a very narrow path, with very little exposure to anything beyond cars and aeroplanes.

In the early '60s, the establishment of coin-operated launderettes had become big business – with substantial profits to match – for owners who chose a lucrative catchment area and looked after their financial interests. Automations International UK Ltd was a franchised offshoot of an identical American company that had been operating in the USA for several years. Its head office was in Hayes, close to Heathrow, with branch offices in Birmingham and Manchester. For sales and service, the country was divided into six operating areas, and I was responsible for the central area, which stretched in a band across the country, roughly between Oxford and Stafford.

With the aid of a salesperson, the client would find a suitable site, usually a high street shop, and negotiate a lease or purchase, subject to a change of use under planning regulations. The company would provide the client with a list of approved installation contractors, who would then offer quotes for the shop conversion and supply of services: gas (or oil), electricity, water, drainage, etc. Firstly, my job was to supervise the installation of the services to ensure that the equipment (washers, dryers, dry cleaners, boilers, and water softeners), which the company would be selling to the client, would work correctly. After opening, the company's service engineers would be on call to carry out regular maintenance and service repairs.

It was good business for everyone concerned and there was a wide range of clients. The average unit took three to four months to complete, from the enquiry stage to opening, depending on the complexity and size of the operation. The job was both challenging and rewarding. No two days were alike, and I travelled substantial distances in my company-provided Cortina estate, whilst meeting a wide variety of people, both good and bad.

After a while though, it became obvious to me that this would not be a job for life. The market would eventually become saturated and competition would grow. After three years, sites were more difficult to find, and the expected opposition had crept in which diluted the market even further. I decided that this was a good time to look for something new. As luck would have it, I had made friends with a man who worked as a sales engineer for a large American company with a UK subsidiary.

Stone-Platt Crawley Ltd was the UK distributors of the range of stone vapour boilers. Its marketplace was industrial rather than commercial and the main attributes of the boilers were their almost instantaneous steam generation, and the relatively small package size for their output. One of the company's most important customers was the Royal Navy and its Merchant Marine, where the boiler was the ideal means of superheated steam generation on board, particularly as a standby. For instance, when in harbour or dock, the main engines would be shut down but steam would still be required for operations such as winching, derricks, hot water generation, and cooking.

An interview was arranged and I was quite surprised to be offered a job, as my knowledge on the subject of steam generation was so limited. However, three months of intensive product training instilled a great amount of confidence and I set off in search of my first customer. This turned out to be the Tyneside shipbuilder Swan Hunter, and it was here that I found myself in the skeletal bowels of a super tanker which was under construction at their Tyneside yard. Sea trials followed my first installation and the world was looking good.

My next customer was Vosper Thorneycroft where our boiler was specified for its range of fast patrol boats. Even here there was a tenuous Jaguar connection, as my first sight of one of these fantastic ships was in Lyme Bay around 1966, when Tommy Sopwith ran his boat, Endeavour, with four Jaguar XK units in the *Daily Express* International Offshore Powerboat Race.

The Huntsman fast patrol boats were provided as guard and rescue ships for this event. I recall one of them, I think named 'Brave Borderer,' lighting up all three of her main engines, and steaming past the flat-out powerboats with majestic ease, bows in the sky, and a huge hole in the sea under her stern from the subsequent thrust; it was magnificent!

Unfortunately, it did not last. After two years both the family and I were tired of the long days away from home and the excessive travelling, so Stone-Platt and I parted company, quite amicably as it turned out, as the parent company in the States soon closed the UK operation down. The problem now was what to do next.

My wife had trained in hotel management and it seemed fitting to use this expertise in some form of business enterprise. As we could not afford to buy a hotel, we decided that a restaurant would be the next best thing. Therefore, in 1972 we sold our house in pretty Southam and bought the leasehold of the Pickwick restaurant in the village of Balsall Common. This was the village where my old friend from the experimental department, Ron Beaty, had his business, Forward Engineering, and was only a short distance from Browns Lane. The significance of this escaped me at the time, but, as will be seen, the Jaguar connection was soon to be renewed.

The restaurant purchase was not one of our most successful enterprises. A combination of Ted Heath's government, the Winter of Discontent (resulting in the three-day week and power shortages), decimalisation, and the introduction of VAT, left our business in a parlous state. We had also discovered that we were not cut out to be restaurateurs! I needed to look for some form of additional income to keep our heads above water, and it was now that Ron came up trumps.

Forward Engineering had obtained the contract to provide the engine, gearbox, axle, and main suspension units for the Bob Jankel designed Panther J72, and as I

A 3.8-litre XK engine, built by Rob Beere, this one in a replica Jaguar E-Type. (Author's collection)

had become the owner of a very tatty Transit van, I took on the job, among other things, of delivering these components to Panther Westwinds.

Bob Jankel's operation was not the most lavish. He operated from his home where he had a drawing office, and the vehicle assembly was a series of lock-up garages in, I think, the Byfleet area. Bob himself was quite a shrewd character and very meticulous in his approach to vehicle construction. The J72 was beautifully built and very much along the lines of a more modern SS Jaguar 100. It started life with the 3.8-litre straight six XK engine, and later was given the 4.2-litre version. Future models even had the V12 engine squeezed under the long bonnet. A large saloon similar to the Bugatti Royale was also produced using the V12 power unit.

The time scale of this operation escapes me, but I suspect that it was about a year before Jankel decided on another supplier.

Meanwhile, Forward Engineering continued to supply competition Jaguar engines to customers who wanted to go racing. Several young engineers served their time with Ron and went on to be very successful in their own right. There was Rob Beere, who would go on to run a well-respected Jaguar tuning business, and David Butcher, a well-known and much sought after Jaguar engine builder.

It was now 1972, and it was around this time that I met John Harper. Ron had rented a corner of his large workshop to John, who at that time was in the business of buying, refurbishing, and selling Lister Jaguars, mostly, if I remember correctly, being returned from America. He was also making a name for himself as a competent racing driver.

John would eventually drive the Forward Engineering E-Type, which at the time was one of the quickest on the club circuits. Working for John, helping to restore the Listers, was a young engineer named John Eales, who, as many people will know, went on to form JE Engineering, specialising very successfully in the further development of the Rover V8 engine for road and competition use. As far as I know, it still does.

During my time at Forward Engineering, I had met Bob Meacham and Bill Pinkney, both of whom were heavily interested in early historic saloon car racing, and both drove Jaguar 2.4-litre Mk Is.

Both of the cars were prepared by Ron, and both became very successful. Bill Pinkney in particular was a very competent driver, and was at the front of the grid most of the time. Ron had agreed to provide support for both cars at the circuits, and I found myself travelling to places like Silverstone, Brands Hatch, Oulton Park, Castle Combe, Donington and Snetterton most weekends, and even to more outlying circuits such as Croft, Pembrey and Valley.

By now our restaurant business really was suffering, and it became obvious that I would need to do more to provide financial support for my family. An old friend from my Jaguar days informed me that the company was now running night shifts on the production lines, and was looking to set on fresh labour. After a great deal of thought and discussion I decided better the devil you know, and made a trip to the employment office at Browns Lane. I attended an interview where I was asked how I would adjust to track life, having spent so much time in the competition and experimental departments. When I explained that this time I was in it for the money and not just job satisfaction, that seemed to be the right answer. Several days later, I received a letter from Jaguar informing me that my application had been successful and that I should report to the employment office the following Monday morning. From there I was directed to the production offices across the link road. I was asked if I needed a guide to get me there and received a strange look when I told them that I could probably find my way blindfolded.

It was a very weird and somewhat comforting feeling to enter the factory for the first time in six years. I was aware, from talking to friends who still worked there, that great changes had been made, but I was totally unprepared for the enormity of them. During the short walk to the offices I met two people that I had known from the past. Both were surprised to see me and wished me luck. I knocked on the door, marked pre-mount production, and when I was called in, there was George Lee, sitting at his desk, barely changed over the years, but now sporting a white cow gown with a blue collar: a senior foreman.

"Bloody hell, Brian Martin, what brings you here?" he said. I handed him my employment slip. After a short pause he looked up and said, in his broad Lancashire accent, "Well I'll be damned. There have been a lot of changes since you left here. I suppose that means I'll have to start training you again."

There was no need to say any more. I knew that I was back home!

Chapter 15

1972-1973

Old friends, assembly lines, and Browns Lane transformed

So there I was, back where I started all those years ago, with little improvement in status (not that there is any importance in that). George took me to the end of pre-mount track number two, and introduced me to the junior foreman of that department, who just happened to be Harry Rudd, an old friend of mine who I'd worked with on the early Mk I line. Now, I would be working for him.

He was quite bemused by the circumstances that he found himself in. "I am supposed to put a man with you to show you how the car, and particularly the wiring, is put together, but, as you were responsible for the design and implementation of most of it, that would almost be an insult."

Eventually we compromised, and I had someone oversee my first operation: installing the main bulkhead harness, multi-plug connectors, and the LH and RH under bonnet forward harnesses. Harry was right; very soon I was waiting for the next body to arrive at my station and was beginning to get bored. Harry and George both decided that rather than introduce me to a different assembly line operation, I would be put straight on to rectification at the end of the line. I quickly learnt that track speed was now god, and that nothing would be allowed to reduce the number of bodies that rolled off the end of the line.

On the third day of my 'new' job, I was busy in the engine bay of the next car when a voice said, "Brian Martin, what the hell are you doing under there?"

I looked up to see a puzzled Peter Craig, dressed in a smart suit, staring at me.

The 4.2-litre Series 2 E-Type OTS. (Courtesy John Starkey/Jaguar Daimler Heritage Trust)

My old friend and original leading hand had now reached the heady heights of plant director and had gained a seat on the board. He realised that there was not much time for explanations, and promptly invited me to his office after my shift, so that we could have a more relaxed conversation. And so, three days in, I was having coffee and biscuits with one of the directors. Peter was very interested to hear about my life and I about his, and our friendly relationship continued until his untimely death, an event which greatly saddened me. God knows what my old union officials would have said about our friendship.

As it happened, I was soon to learn that they hadn't stood still either. My old shop steward, Walter Turner, was now plant manager, and the original NUVB convener, Harry Adey, was personnel manager.

I suppose, thinking about it, the attributes of a successful union official are very similar to those of a manager: good communication skills, the ability to compromise, and an understanding of procedure. On top of that, of course, from senior management's point of view, what better way to remove a potential adversary than to persuade him to change sides? Quite a shrewd move.

I had to rejoin the union (NUVB) as my membership had long since lapsed.

Jaguar was not a total closed shop, but, in practice, it would have been impossible to work without a Union card.

I now had time to reassess my surroundings. The Browns Lane plant had changed beyond recognition. For starters, it was now referred to as the British Leyland Large Vehicle Plant, since Jaguar had, by this point, been absorbed into the British Leyland (BL) conglomerate. Engine assembly and sub-components had now moved to the Daimler factory in Radford. Complete front and rear suspensions were also being assembled at this factory. These major components were then being delivered to Browns Lane, which was now just an assembly plant, ready to be installed in the bodyshell on the new mounting tracks.

The machine shop had also been moved to Radford. The brass and chrome shops, and all of the small departments that had made Browns Lane a complete, independent unit, were gone as well. The main assembly halls one and two had been gutted. In the space once occupied by the machine shop and engine assembly, were new production lines for pre-mount, mounting, and trim. The old tin sheds that had separated the main factory from the sports field had been torn down, the area roofed over to form a new goods-in department, and an exit for finished cars to go out on road test. The Series 3 E-Type, now in the middle of its lifespan, had its own assembly line.

The Series 1 XJ was about to go out of production and a new model, the Series 2, was being introduced. For a period these two ran together, with long wheelbase versions of the Series 2 mixed in with the short wheelbase cars.

As already mentioned, a two-shift production system had been introduced by the British Leyland board to increase production numbers, who seemed determined to erase the name of Jaguar from the history books. The final sacrilege was the removal of the 'leaper' badge from the lower front wings, a nasty blue 'L' put in its place. History shows that this period set Jaguar down a slippery slope which would lead to its loss of identity, and its eventual demise as an individual manufacturer.

Settled into my job, I was given the last two positions on the assembly line, just before the body was lifted and moved by conveyor to the mounting track, where it was mated to the front and rear suspensions, the engine, and the gearbox. Jaguar had continued with its tried and trusted method of inspection.

An early XJ6.
(Courtesy John Starkey/Jaguar Daimler Heritage Trust)

A dedicated information and inspection manual accompanied every car from start to finish. At specific points during the assembly process, the works previously carried out would be inspected and, in theory, signed off by the inspector. In practice that often did not happen. This system had worked well in the early days of the Mk I, II, VII, VIII and IX, and the E-Type, however, the XJ was a much more complicated beast, made even more so by the introduction of air conditioning, and the use of untrained labour.

Air conditioning had once been an optional extra but it was now more or less a standard fitting. It was a heavy and complicated unit requiring two operators to install. It came to the track as a complete sub-assembly, and could only be tested much further down the line, after gassing and electrical power had been installed. This meant that any rectification, and there was a lot of it, would require a complete strip down and replacement of the unit. As an average this would take two operators in excess of two hours to complete, off line. There was little allowance for rectification work immediately after the inspection process, without stopping the assembly line, which, remember, was god! Still, BL pushed for a higher volume of cars, desperate to raise its profits.

The result was faults on top of faults, and more rectification work was carried out, off line and often during overtime, on cars that were ninety per cent complete. This did nothing for Jaguar's build quality, and customer perception of the marque began to slide very quickly, a far cry from the heady days of the fifties and sixties. Meanwhile, BL had problems of its own. Lack of investment in plants and machinery, antiquated design, poor quality control, even poorer industrial relations, and over production led to huge stockpiles of finished vehicles that no one wanted, and few could afford anyway. These were all signs of the times for the British motor industry.

This is what led to Jaguar's introduction of a production night shift in an attempt to increase volume and thus profit. The day shift would stop at around 5.00pm, and the night shift would start at around 7.30pm, running through to about 6.00am. The gaps in between were used for re-stocking and maintenance. All workers were expected to take on night shifts, and the shift pattern changed every two weeks. Some workers, however, if it suited them to stay on the night shift permanently, were allowed to do so, provided that their opposite was happy to run the day shift. It worked reasonably well, but only resulted in even more complete and unsold vehicles being parked on the sports field, the car parks, and several redundant airfields around the Midlands. Sir William's ethos of not building a car unless it was already sold had long since gone! In 1972, he, the guiding light of Jaguar ever since its inception, had retired aged 71. His place was taken by Lofty England, the most unlikely person to ever run a car factory.

One of the greatest problems that Jaguar had at this time was paint quality and paint shop capacity. The equipment was very old, some of it going back to the Swallow Road days. Sir William was never very good at spending money; remember the pre-mount assembly line, bought second-hand from the Mulliner factory in 1953?

The introduction of higher quality metallic and non-solvent based paints became difficult to handle. A new, larger, and more modern paint shop was needed, as was the space occupied by the current one. There was no question, it would have to go elsewhere. Enter Geoffrey Robinson.

Mr Robinson was a Labour MP for the Coventry North constituency and a member of the current Labour government. He also had close ties with the motor industry through his time as CEO for the Italian motor manufacturing company, Innocenti, which, among other things, produced a re-styled Italian version of the original Mini.

The Labour government had taken a stake in the failing BL group and, in particular, a golden share in Jaguar. There had been rumblings in the marketplace that Jaguar could be sold off, and it was well-known that both Ford and General Motors were interested. The golden share was intended to avoid such circumstances until the government felt the time right. Meanwhile, the government needed a socialist executive (Geoffrey Robinson) to guide at least one part of the uncontrollable monolith BL had become, back to safe profitability.

By this time, it was 1973. I had now become a foreman and controlled a third of the long pre-mount track. One of Robinson's first actions was to call a meeting of all production staff. We were entertained by a lecture on how he and the government saw the shining future of Jaguar under his stewardship. Needless to say, most of us were quite quickly convinced that neither he, nor anyone else in Whitehall, had the remotest idea of how to run a manufacturing plant, so the future did not look good.

Robinson laid out their ambitious plans for the expansion of Browns Lane at the meeting. They would purchase the wedge of land bordered by Browns Lane and Brownshill Green Road. A new manufacturing facility would be built to house a modern paint shop and other supporting units, and would be surrounded by a vehicle test track to enable production cars to be road tested without using public roads. To support the necessary planning application, a new public access road would be built through the Counden Wedge, and off it a new rear entrance to the Jaguar plant. All Jaguar traffic would use this entrance, dramatically reducing the traffic flow on Browns Lane. This would, they hoped, mollify the local residents whose objections to the whole proposal were getting pretty voluble.

As it turned out, it was all a damp squib. Sometime later the residents had their new road, Jaguar its new entrance, and the rest of the scheme was cancelled. Although not before our illustrious leader had arranged the purchase of a few hundred thousand pounds of steelwork from Italy, intended for use in the construction of the new paint shop. This was duly delivered and deposited on the sports field, where it lay rusting away for many years, before finally being sold for scrap. It was nicknamed Robinson's Clanger, and for some time carried a sign to that effect.

To be fair to Robinson, the location of the new paint shop was taken out of his hands by the Leyland board who, following the Ryder Report, decided that it should be located at the Castle Bromwich plant. This meant that all Jaguar bodyshells, alongside other models produced by the group, were painted there and shipped in specially designed trucks to Browns Lane.

Robinson was particularly close to the unions with his Labour Party roots, and this created massive problems for junior management. It was almost impossible to prescribe any form of criticism to a line worker without cries of victimisation, and calls for a senior management meeting to discuss the matter.

The very last E-Type built, a Series 3 OTS, with a V12 engine, in the JDHT collection. (Courtesy John Starkey/Jaguar Daimler Heritage Trust)

Senior management were weak; the union representatives held the cards and any compromise was usually in their favour. We were trying to run a modern production unit with our hands tied behind our backs.

There is one particular moment that sticks in my mind. It was the last shift before the Christmas break, a night shift, and Robinson came down the line just before midnight, shaking the hands of every operator, and wishing them a merry Christmas and a happy new year. At the same time, he walked past every foreman and ignored them.

Christmas of 1973 was tough at home. My wife and I were still trying to run our non profit-making restaurant, she doing most of the work while bringing up our two girls. It was hard, and put her off restaurants for life! Very soon, it became too much and we decided to sell. We managed to find a buyer at our original purchase price, which, in hindsight, we were very lucky to do. The Pickwick, as it was known up until then, became an Italian restaurant called La Pergola, practically overnight, before later becoming an Asian restaurant called Café Tamerind.

Chapter 16

1974-1978

The XJC, the XJS, and leaving Jaguar

Using the proceeds of the sale of the restaurant, we put down a deposit on a small house, less than a mile from the restaurant, as the children's schools had already been decided. In my work life, Jaguar was not finished with me yet, as this final chapter will explain.

The Ryder Report of 1974 saw more nails in the Jaguar coffin. By this point, Lofty England, who was never cut out to run a car factory, had retired and was happy to go. The post of chief executive at Jaguar was appointed to chief engineer, Bob Knight. This meant he also became the Jaguar representative on the Leyland board. This was probably a good thing for Jaguar because Bob Knight, bless him, had never been known to make a decision in his life, and was forever making long speeches, which must have bored the other members to death, leaving him, to some extent, to his own devices. These were important times for Jaguar, especially when it came to retaining some semblance of independence.

Early in 1975, Geoffrey Robinson resigned, and left to pursue his political career. Soon, Jaguar would introduce its first new model for many years. Most people thought it would be the long-awaited Jaguar sports car, the replacement for the beloved E-Type.

Previously, in 1973, a new model had been demonstrated to the press. It was the XJ-C, a two-door coupé, which was an elegant car. Its most striking feature was the elimination of the 'B' posts, which created a pillar-less design. Without these posts, a clear area of glass, with no obstruction, was achieved; the electrically operated quarter-lights met the door glass where the 'B' post would normally have been.

It was based on the short wheelbase version of the XJ platform, and used the front and rear of the then in production Series 2 saloon. Despite this introduction it would be two years before the car came into serious production. There were many problems associated with the production of this model, not the least of which was the operation and sealing of the complicated rear quarter-lights. The ingress of water could result in catastrophic failure of the rear seat pan, as many owners would find out. Only 6505 XJ6Cs and 1873 XJ12Cs would be produced, before the model was dropped in favour of, and to avoid clashing with, the XJS.

This, the XJS, was the long-awaited and much discussed replacement for the E-Type. It was unpopular from the start, both inside and outside the factory. Many had expected to see a modern version of the E-Type, commonly referred to as the F-Type. Chief designer, Malcolm Sayer had always intended his new car to be a grand tourer, not a sports car. Based on the saloon platform, with large overhangs front and rear, it was a very large motor car, which rather irreverently earned it the nickname 'the barge.' It would be many years before the XJS was accepted as a Jaguar classic, but now it has a worldwide cult following. I myself have owned two, and my wife and I have spent many happy times touring in comfort through France, Spain, Germany, and Italy.

The XJS had a very difficult gestation period. It was built on Jaguar's new

A 1976 5.3-litre Jaguar XJ Coupé. (Courtesy John Starkey)

Me touring in my XJS-C convertible. (Author's collection)

caterpillar assembly line, which, for the first six months of production, spent more time stationary than going forward.

During this time, and against much advice, Leyland had introduced a thermoplastic acrylic (TPA) finish to its new paint facility. This had been tried in the USA and failed dismally. There were massive problems from the start, with the colour finish refusing to adhere to the primer undercoat, and I personally witnessed cars that had returned from their first road test with sheets of the outer finish stripped off by the wind. The problem was eventually solved, but not before a substantial number of cars were stripped back to bare metal and repainted at Browns Lane. There were also component shortages and design upgrades to contend with, all of which delayed production, and in turn delivery to the showrooms of Jaguar dealers.

By 1978 morale at Jaguar was at an all-time low and there was talk of closure. The story of how Jaguar clawed its way back from this abyss has been well recorded by others, but for me it was time to review my future. I had been approached by a friend of mine to join him in a venture, which

The old Browns Lane factory being dismantled. (Author's collection)

had nothing to do with motor cars, and much to do with the production of thermoplastic materials.

In June of 1978, I left Jaguar for the second and final time. It was a sad occasion; the company had been part of my life for so many years and there were so many memories to recall. Time marched on, however, and new pastures called.

Still, my relationship with Jaguar proved impossible to break, and in 1997, after my retirement, I returned to Browns Lane as a volunteer in the Jaguar Daimler Heritage Trust museum. There, I spent the next five years in the company of the cars that, in my own small way, I had helped to design and build.

Now, it has all gone. In the space where the Browns Lane factory stood is a housing estate. The museum is no more and the old cars are scattered across other sites. All that remains are ghosts and memories.

Epilogue

My Jaguars

Ever since I bought my first Jaguar in 1957, I have rarely been without one in my garage. This epilogue contains a list of them all.

1957
A 1951 2.5-litre Mk V DHC

A 2.5-litre Mk V DHC, similar to the one I bought in 1957. (Courtesy John Starkey/Jaguar Daimler Heritage Trust)

The Mk V was black with a red leather interior, and a black mohair hood. I purchased it in 1957, using the money I had saved whilst in the RAF and my demob fund. I really should have spent it on my family, but, when I saw her languishing on a used car lot in Nuneaton, the temptation was too great, and for that I sincerely apologise to them. I reluctantly accepted the asking price, and parted with £240, becoming the proud owner of my first Jaguar!

She had been used as a wedding car and for that reason was immaculate, both inside and out. Mechanically, though, it was a different story. What seemed to be a simple misfire turned out to be a failed head gasket between two of the six cylinders, and removing the head revealed the need for a long overdue decoke.

Once that was done, and after fitting the cleaned up valves, springs, and new valve guides, the engine ran much better. I drove the Mk V for just over a year, including two weddings, which helped towards the astronomical fuel costs. Whilst still outwardly immaculate, rust was creeping insidiously into the bulkhead and boot, and unless I could take action to stem the tide, I would end up with a large bill for repairs.

I did not have a garage or the necessary equipment to take on that sort of work, so she had to go. I managed to sell the Mk V to a local wedding car hire company and made £100 of profit towards my next Jaguar.

1961
A 1954 3.4-litre XK120 FHC

An XK120 FHC, similar to the one I owned in 1961. (Author's collection)

Suede green with matching leather and lots of rust, the XK was a very sorry-looking lady. She'd had several owners, had very high mileage, a rough running engine, and no history. The asking price was £210, which I eventually haggled down to £175. The owner later told me that he was just relieved to get rid of it. It was obvious from the start that only five of the six cylinders were in use.

By now I had made friends with a local garage owner, who, like me, was a member of the NMC. For a nominal rent and a contribution to the electricity bill, I had the run of his garage after hours.

Using these facilities, I removed the cylinder head. To my dismay, the exhaust valve for cylinder number five was minus a head and there was neat hole in the piston where it had gone through. Now the sump had to come off, and it was all beginning to look expensive. Fortunately, the valve head had gone straight through the piston and the bore was undamaged. Removing the sump, I found the errant valve head embedded in a glutinous layer of black sludge. I took out the piston and con rod and replaced them with new (scrapyard new) ones. I even managed to rescue the original piston rings! With an exhaust valve, new timing chain tensioners, head gasket, oil filter, and fresh oil the car was back on six cylinders. All that remained was to drive the car to the experimental car park at Browns Lane and get the wizard of SU, Stanley Woods, to reset the carburettors. The engine now ran perfectly, unlike the rest of the car.

The braking performance was appalling; the car pulled heavily to the left and made no effort to stop. I went to Jaguar's salvage department and found a pair of Dunlop discs and calipers that had been discarded from one of the test cars. They were just what I needed and I paid £20 for them. Fitting them to the car took a lot of thought, and required the fabrication of caliper mounting brackets, but eventually I was satisfied.

When I removed the rear drums, I found a great deal of oil where it should not be courtesy of the failed and worn axle tube seals. The positive result of this was that there was no rust in the workings. After a thorough clean of everything, the oil seals replaced, a set of new Mintex brake linings, and the diff cleaned out and filled with fresh Castrol EP90, I had brakes! The car pulled up cleanly in a straight line and retardation was much improved.

The paintwork came up quite well with a cut and polish, as did the interior, but although I touched up the rust spots, it really needed a strip back to bare metal and a respray, which I could not afford. I had also spotted my next Jaguar so it was time for the XK to go.

1961
A 1956 2.4-litre Mk I

A 1956 2.4-litre Mk I. (Courtesy John Starkey/Jaguar Daimler Heritage Trust)

I advertised the XK for £750 which I thought was a fair price considering the time, effort, and cash that I had put into it. It was obviously under-priced, though, as the first viewer bought the car, paying me in cash!

The car that I had seen advertised in the Coventry Evening Telegraph was a 2.4-litre Mk I saloon, manual transmission with overdrive, BRG paint with tan trim. It was only five years old, had low mileage, and one previous owner, which all meant that it had been well looked after. I negotiated a price of £510 and became the proud owner of WNX 851. It was a much more practical vehicle for my young family, with a fuel consumption that I could afford. I was now a member of the JAMC and tended to compete in most of their events. The Mk I was not really suitable for sprints and driving tests, but I tried and was rarely embarrassed.

The reader will have gathered by now that I could never leave my cars alone, and the Mk I was no exception. In search for more power, I consulted Stanley Woods. His immediate response was to get rid of the air-strangling Solex downdraught carburettors, and fit a decent pair of sidedraught SUs instead. There was no suitable manifold for this conversion, so I persuaded Bob Blake, ace welder and fabricator of the competition department, to chop and weld a 3.4 manifold at an inclined angle to clear the inner wing. A pair of SU HD6s, tuned by Stanley, completed the job. With this modification and further work on the exhaust manifold, I reckon we gained a further 20-30 bhp and an increase in torque.

I ran the Mk I for about four enjoyable years before having to part with her for domestic reasons.

1978
Two 3.4-litre Mk IIs

*A 3.4-litre Mk II.
(Author's collection)*

After many years with mundane company cars, I was ready for my next Jaguar. In fact this time it was two! A friend of mine from the RAF found himself needing to sell his Jaguar fleet, which included two Mk IIs. The first was a 1959 3.4-litre, manual overdrive, BRG with tan trim, registered COV 38. She was in a sorry state but I purchased her as a donor for the second car. This was a 1960 3.8-litre, manual overdrive, white with red trim, registered 111 LOO.

The total cost was just over £600. Over a period of three months I stripped COV 38 of anything that I might be able to reuse or sell. The engine and gearbox, rear axle, and front suspension sold for just over £300, thus already halving my initial costs. The woodwork, instruments, and many of the small components went into store. The remains (complete with the log book and registration number) were dragged on to a low-loader and taken to a scrapyard in Birmingham, making me £50. It was not until sometime later that I realised the significance of COV 38; Coventry 38 would have been the perfect registration for my replica XKSS, which had a 3.8-litre engine. That was one that got away!

I made sure to not make the same mistake with 111 LOO. I still have that registration in my possession and it has been the number of every Jaguar I have owned since. I ran the white car as everyday transport, including two trips to Spain. During one of these, whilst on the autoroute just outside Bordeaux, my eldest daughter, Charlotte, complained of a nasty smell in the back of the car. Looking in the mirror, I was horrified to see that I was being followed by a large cloud of white smoke. I pulled over as soon as possible and went to investigate. I found the whole rear of the car covered in a thin film of evil smelling oil.

It turned out that I had made a fatal mistake when refurbishing the rear axle. I had painted over the vent hole in the differential cover plate, so the diff had pressurised and blown the nose seal. The oil had leaked on to the exhaust system, hence the white smoke! I could do nothing to repair the seal. All that I could do was clear the vent hole and find some axle oil to top up the diff. If you have ever tried to find EP90 oil in front-wheel drive France, on a Sunday no less, you will understand my dilemma!

After my fourth failed attempt to use my basic French, and a lot of driving very slowly whilst praying that there was still some oil left in the diff, I finally found a rundown old garage in a small rundown old village. The ancient proprietor spoke no English (why would he) but, after persuading him to lie under the back of the car, I pointed at the diff, and the oil on the car, and an air of understanding crept over his face. "Un moment monsieur," he said before he rushed back into the garage, reappearing a minute later, proudly waving a litre of Castrol EP90. "Deux s'il vous plait," I replied, and he brought me another, plus a small funnel with a length of plastic tube.

I do not remember how much he charged me, but I would gladly have paid in gold for those two bottles of oil. I was even more impressed when the garage doors were opened, and he gestured for me to drive the car on to an ancient four poster ramp. I was ushered away whilst he topped up the diff. What a star! He definitely did his bit for *entente cordiale*. We then continued our drive to L'Estartit at a reduced pace.

I topped up the diff three times during that holiday, and on our return, replaced the nose seal. I was very lucky that there had always been enough oil left to protect the final drive unit.

1985
A 1966 2.8-litre Series 1 XJ6

Sandy. (Author's collection)

We had a lot of fun with the Mk II as a family, but I had spent a great deal of time at Jaguar on the development of the first model in the XJ series, and I wanted one! After parting with the Mk II and £520, I became the proud owner of 'Sandy.' Named by my daughter Charlotte, because of the car's Green Sand paintwork, Sandy was the first of my Jaguars to have a name. She was a 1966 2.8-litre Series 1 XJ6, manual overdrive, with tan trim. She was in a beautiful condition, and for the first time, I had a Jaguar that only needed cleaning and polishing!

There was a great deal of scaremongering about the short block XK unit. Piston failure could easily occur in a part-worn, heavily coked car, especially if it was suddenly driven hard for a long period on low grade fuel. The resultant rise in temperature was enough to melt a piston in some cases. The higher piston speed in the short block configuration, when compared with the 4.2-litre, did not help. The 2.8-litre was produced as a cheaper alternative to its bigger sister, particularly in France, where there was a punitive tax on engines over three litres.

We did many miles in Sandy, including two continental holidays to France and Spain, a trip to Germany for the Nürburgring Oldtimer event, as well as many Jaguar events in England, and she never missed a beat. All that she ever asked for was fuel, occasional oil, to be kept clean, and be tucked up in her garage at night.

She did eventually develop a definite growl and a whine on the over-run from the back end, which could only mean one thing; a problem with the differential. As I now had other everyday transport, I could afford to take the XJ off the road, and rather than mess around under the car, I took the whole suspension out. On inspection, both crown wheel and pinion were showing excessive wear and free play, but the bearings were fine. With these replaced, new discs and pads, and four new shock absorbers, she ran like a dream again.

On September 29, 1996, after suffering severe chest pains for several days, I was admitted to hospital and diagnosed with severe blockages in the arteries that feed the heart. This required a quadruple by-pass operation to correct. Seven days later I was back home, with instructions from the specialist to walk at least two miles a day. Within a month, I was back at my desk, and, touch wood, I have not had a problem since, although I do take a meal of tablets everyday. All of this promoted a reappraisal of my life; I had always wanted to build a car from scratch, so I decided to build a replica!

1999
A replica XKSS

My replica XKSS. (Author's collection)

Having made that decision, I looked around at what was available, and I quickly shortlisted two: an AC Cobra (a car that I much admired), or a Jaguar XKSS. As it happened, LR Roadsters, who were then situated just outside Cambridge, produced kits for both models using the same chassis, designed by Adrian Reynard. After a great deal of thought I decided to go ahead with, naturally, the XKSS.

One day in October 1997, I borrowed a trailer and hotfooted it to their door to collect the basics: chassis, body tub, and panels.

I had already purchased a donor car, a sad-looking Series 2 XJ. I spent the next three months (evenings and weekends) in the garage, stripping and refurbishing the components and slowly building them into the chassis. Then, once I had fitted a set of slave wheels (off the XJ), I could mount the tub, which was held to the chassis by four long bolts. I had purchased a pair of seats, a hood, replica D-Type wheels, and a set of triple 45 DCOE Weber carburettors. After fitting the bonnet, doors, and spare wheel lid, it was time for paint. Now in less pieces than she was before, Sylvia (so named because she was to be painted silver) went back on the trailer and off to the paint shop. As it turned out, last minute decisions meant a change of plan, and she was painted metallic mica flamenco red, which turned out to be a much wiser choice. However, the name stuck.

The day came for the first engine run. After some minor adjustments to the brakes and the carburettors, I was ready for a road test and many miles of pleasure. The build took a year and a half to complete, and this is just a snippet of what I'm sure could fill another book.

In order to fund the XKSS, I had sold Sandy. There were tears from the family when she went, but it was for a good cause. We had some equally good times with Sylvia: three continental tours where she was much admired, two trips to 24 Hours of Le Mans, and a place on the Innes Ireland Memorial Run to Silverstone, where she was driven by none other than Richard Noble!

2002
A 4-litre XJ40

Now that I could afford to replace Sandy, I purchased Blue. She was a 4-litre XJ40, painted in light blue metallic, with blue trim, and in pretty good condition. I ran the XJ as an everyday car, commuting to Wolverhampton, where my company was now based, for over a year. Despite the negativity surrounding the XJ40s, I have to say that she never let me down. Then one day, I was at the garage where our cars were serviced and I saw Lady; I fell in love.

2003
A 1995 4-litre X300

My X300. (Author's collection)

Lady was an X300 in British racing green with biscuit trim, a wonderful 4-litre AJ6 straight-six engine, which in my opinion is the best engine that Jaguar ever designed, and good for over 200,000 miles when looked after.

Registered in 1995, she had only completed 65,000 miles and was in pretty good condition. So it was Blue's time to go. Fortunately, I found someone local to Wolverhampton who took her off my hands.

The X300 is arguably Jaguar's finest saloon car. It may not have the charisma of a Mk II or a Series 1 XJ, but it was competent, with a magnificent ride, and that silky smooth straight-six, but I was biased.

My wife and I did several continental trips in Lady, including Italy, to visit the Ferrari museum at Modena, and Germany, for the historic 6 Hours of Nürburgring.

In 2007, we made the momentous decision to move to France. We bought a beautiful old house in a village some 50km north of the foothills of the Pyrenees. However, before we left, I exchanged Lady, in a straight swap, for Gracie.

2007
A 3.6-litre
XJS-C

My XJS-C.
(Author's collection)

Gracie was an XJS-C cabriolet in regent grey with oatmeal trim, and the much sought-after factory hard top. She had the 3.6-litre AJ6 straight-six engine, and the manual 5-speed Gertrag gearbox.

We drove Gracie to France on English plates, and then spent a month and £300 wading through the French bureaucracy to get her registered on French plates and insurance. We built a garage for her in the barn of 'Maison Blanche' (named after the very well-known spot on the original Le Mans circuit), and tucked her away for the winter.

In 2008, we completed a tour of Northern Italy, taking in the Alps and some of the well-known Alpine passes. We also took part in several Jaguar events, including Le Mans in 2009.

My world fell apart in November 2011, when, after a short illness, my wife passed away. I had no intention of living in France on my own, so the house went up for sale. It sold well, and in December 2012 I was able to move back to England, taking Gracie, and my wife's ashes, home.

I moved in with our eldest daughter, Charlotte, and her partner, Gary, in the village of Byfield, midway between Banbury and Daventry. There, to house Gracie, we had a seven-metre square garage built.

With my health slowly deteriorating, Gracie was not being used. In 2016, I very reluctantly decided that she should go to a new home, where she would be used and cared for. Only two people responded to my advertisement. The first one, I am glad to say, thought the price too high. It turned out that he was a private dealer who only wanted to make a profit by selling the car on.

The second, a German gentleman, agreed to a price close to what I was asking. Shortly after his first viewing and road test, he turned up with a trailer and Gracie was on her way to her new home in Germany.

2017
An XJ8 (X308)

My X308.
(Author's collection)

You may assume that my life with Jaguar had come to an end, but that was not the case. I saw an advert online for an X308, which, as enthusiasts will know, is the V8 engine version of X300.

The X308 was at a local dealer for silly money, as he just wanted to move it on. It was love at first sight, metallic emerald green with ivory trim. Esme (named again by my daughter) had covered 136,000 miles, and I parted with less than £2,000 to claim her. Despite her high mileage, and the almost certain fact that the dreaded tin worm would be lurking under those awful chrome wheelarch covers, she was a lot of car for the money. Those covers have gone, and the tin worm has been despatched. The dodgy, leaking power steering pump has been replaced, and she now sits in the garage waiting to be used. At the moment, I am happy to just go in and look at her!

Is she the last of my Jaguars? Who knows.

Also from Veloce Publishing –

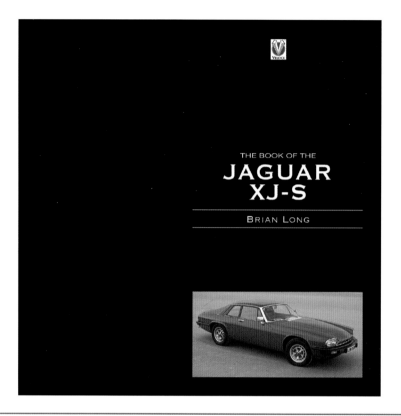

New large format edition of the definitive history of Jaguar's E-Type replacement, the XJ-S. More a grand tourer than a sportscar, the controversially styled XJ-S offered a combination of supercar performance and grand tourer luxury. Includes rare photos of the prototypes that didn't make production.

ISBN: 978-1-845844-01-1
Hardback • 25x25cm • 160 pages • 290 colour pictures

For more information and price details, visit our website at www.veloce.co.uk
email: info@veloce.co.uk • Tel: +44(0)1305 260068

Also from Veloce Publishing –

This significantly enhanced Fourth Edition of Jaguar – All the Cars, brings the Jaguar model story right up-to-date. The only publication available covering the entire range in precise detail, with a revised engine chapter, updated chapters on existing models, and new chapters on the very latest Jaguar models.

ISBN: 978-1-845848-10-1
Hardback • 23.5x17cm • 392 pages • 700 colour and b&w pictures

For more information and price details, visit our website at www.veloce.co.uk
email: info@veloce.co.uk • Tel: +44(0)1305 260068

Also from Veloce Publishing –

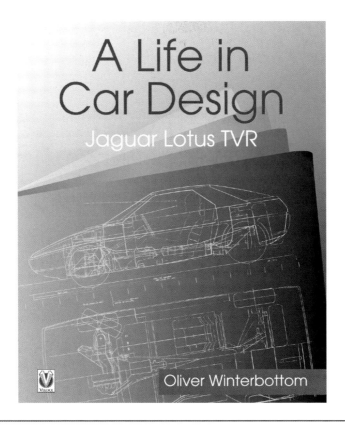

This book gives a unique insight into design and project work for a number of companies in the motor industry. It is aimed at both automobile enthusiasts and to encourage upcoming generations to consider a career in the creative field.
Written in historical order, it traces the changes in the car design process over nearly 50 years.

ISBN: 978-1-787110-35-9
Hardback • 25x20.7cm • 176 pages • 200 pictures

For more information and price details, visit our website at www.veloce.co.uk
email: info@veloce.co.uk • Tel: +44(0)1305 260068

Also from Veloce Publishing –

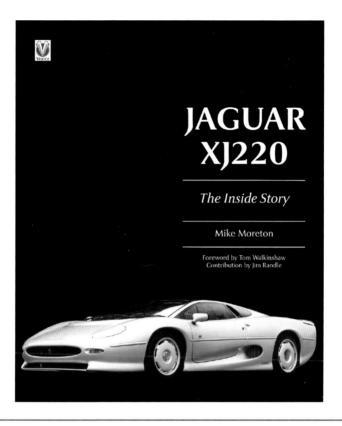

The Jaguar XJ220 supercar evolved from Jim Randle's sensational 1988 UK Motor Show concept car.
The planned production of 350 limited edition cars, each with a price tag of £360,000, was over-subscribed by a factor of four in a single day!
In this book, Mike Moreton, ace director of impossible projects, who was headhunted for the XJ220 project by Tom Walkinshaw, relives the extraordinary inside story of this fantastic, hi-tech car.

ISBN: 978-1-845842-50-5
Hardback • 25x20.7cm • 160 pages • 225 colour and b&w pictures

For more information and price details, visit our website at www.veloce.co.uk
email: info@veloce.co.uk • Tel: +44(0)1305 260068

Also from Veloce Publishing –

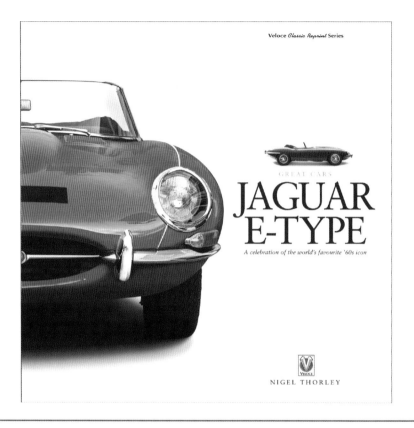

This book provides an insight into the background, development and features of Jaguar's iconic E-Type in both historic and modern day terms, and is beautifully illustrated with superb archive images and modern studio photography.

ISBN: 978-1-787110-25-0
Hardback • 25x25cm • 184 pages • 264 pictures

For more information and price details, visit our website at www.veloce.co.uk
email: info@veloce.co.uk • Tel: +44(0)1305 260068

Also from Veloce Publishing –

The compact Jaguar saloons of the '50s and '60s epitomised the image and resulting success of the Jaguar marque. This book provides an insight into the background, development and features of the Mark 1 and 2 cars, and is beautifully illustrated with superb archive images and modern studio photography.

ISBN: 978-1-787110-24-3
Hardback • 24.8x24.8cm • 160 pages • 269 pictures

For more information and price details, visit our website at www.veloce.co.uk
email: info@veloce.co.uk • Tel: +44(0)1305 260068

Index

A

Adey, Harry 36, 163
Alvis 11
Anglia, Ford 146, 153-155
Apprentice(s) 11, 13, 27, 79, 80-82, 85, 88, 90-92, 96, 125, 127, 151, 152
Apprenticeship 10 ,11, 27, 78, 85
Aston Martin 11, 58, 63, 129, 133, 137
ATC 10, 38
Austin 8, 131, 143

B

Bacon, Sam 90, 91
Baker, Shaun 124, 125
Beaty, Ron 77, 78, 80, 82, 85, 110, 117, 138, 159
Bike 8, 9, 11, 17, 26, 27, 65, 89, 90
Blackpool 11, 14, 16, 48, 53
Blake, Bob 82, 103, 116, 132, 133, 140, 151, 176
Boden, Terry 110, 135
Brooks, Ted 20, 80, 82, 147
Browns Lane 1, 3-6, 49, 52-54, 58, 71, 78, 83, 84, 90, 102,107, 109, 113, 117, 119, 121, 123, 124, 126, 129, 139, 142, 146-149, 159, 161, 162, 164, 166, 167, 171, 172, 174
Buck, George 79-81, 127

C

C-Type 63, 65
Cassidy, Bill 77, 79, 80, 87, 88, 114, 115
Competition 2, 4, 6, 9, 24, 37, 62, 79, 80, 82, 83, 87, 90, 91, 101-103, 105, 107, 109, 114, 131-133, 136, 137, 140-142, 147-152, 155, 156, 158, 160, 161, 176
Coombs, John 102, 131, 133, 137, 140, 141, 148
Coventry 2, 9, 10, 12, 15, 33-35, 39, 45, 51, 64, 65, 74, 81, 82, 88, 91, 116, 119, 121, 123, 127, 130, 138, 144, 155, 166, 175, 177
Craig, Peter 16-18, 23, 26, 27, 36, 38, 53, 56, 67, 76, 77, 163
Crewe 11
Cunningham, Briggs 108, 109, 136, 137, 141, 142, 147

D

D-Type 82, 105
Daimler 6, 15, 21, 32, 35-37, 53, 59, 61, 65, 66, 68, 75, 76, 88, 99, 105, 108, 113, 120-124, 127-129, 131, 145, 148, 150, 163, 164, 167, 171, 173, 175
Derailleur 11
Dewis, Norman 79, 84, 98, 99, 104, 109, 123, 126
Dunlop 11-15, 24, 27, 62, 73, 88, 106, 109, 175

E

E-Type 84, 87, 94-96, 99, 103-111, 113, 116, 117, 120, 121, 123-126, 128, 131, 134, 136-138, 140, 142,146, 151, 155, 159, 160, 163, 165, 167, 169, 170, 179, 184
E1A 4, 101, 105, 107, 108
E2A 4, 82, 99, 101, 107-109, 136
Earls Court Motor Show 20, 30, 31, 81
Eastick, Jim 80, 82, 85
Ecurie Ecosse 58, 65, 69, 82
Electrical 17, 41, 45, 55, 72, 77, 78, 80-82, 84, 86, 101-103, 106, 109, 110, 114, 127, 131, 132, 135, 137, 141, 147, 149-152, 165, 170
Electricity 9, 41, 157, 174
Engineering 5, 9, 10, 25, 27, 39, 77, 78, 83-85, 101, 104, 105, 110, 114, 124, 128, 131, 132, 136, 141, 152, 155, 157, 159, 160
England, 'Lofty' 22, 66, 73, 86-87, 91, 107, 111, 165, 169
Experimental 4, 6, 14, 22, 62, 77, 79-83, 86-88, 95, 109, 110, 113, 114, 116,

117, 119, 121, 123, 128, 139, 147, 154, 155, 159, 161, 174

F

Fangio 57, 58
Farndon, Bertie 26-28, 34
Fenton, Alice 11, 13
Ferrari 57, 61, 62, 72, 133, 137, 140, 146, 181
Foleshill 15, 21, 28
Ford 4, 15, 44, 45, 51, 52, 56, 57, 79, 87, 92, 93, 95, 107, 112, 123, 131, 139, 154, 155, 157, 164, 166
Forward Engineering 77, 159, 160

G

Gannon, Jack 79, 80
Garage 9, 56, 70, 87, 90, 121, 124, 125, 141, 143, 149, 154, 160, 173, 174, 177-183
Gardner, Fred 22, 23, 28, 80, 110
GEC 79, 89, 113, 114, 143, 146
GTO 112, 133, 137, 140, 146

H

Hawthorn, Mike 57, 58, 65, 96, 107, 117
Haynes, Bill 84, 88, 105, 107, 151, 152
History 6, 7, 10, 14, 39, 53, 72, 82, 123, 124, 164, 174, 184

I

Ireland, Innes 111, 112, 133, 137, 180

J

Jabbeke 20, 68
Jaguar Car Club 6, 11, 22, 63, 88, 90-92, 122, 143, 146, 149, 153, 156, 160
Jaguar Daimler Heritage Trust 6, 16, 21, 32, 35, 36, 53, 59, 61, 65, 66, 68, 75, 76, 99, 105, 108, 113, 120-122, 127, 129, 145, 148, 150, 163, 164, 167, 171, 173, 175
Jaguar Drivers Club 7, 90, 91, 156
Jaguar Enthusiasts Club 6
Jankel, Bob 159, 160

K

Knight, Bob 64, 79, 80, 87-89, 115

L

Le Mans 2, 4, 8, 11, 49, 52, 56-58, 61, 62, 65, 69, 81, 82, 86, 103, 104, 107, 108, 117, 136, 137, 142, 146-148, 180-182
Lee Enfield .303 9, 40
Lee, George 16, 19, 27, 30, 53, 67, 77, 161
Leeming 41-43, 46, 48, 49
Leeson, Percy 31
Leyland, British 25, 156, 164, 167, 169, 171
Lucas 17, 77, 83, 84, 102, 104, 110, 136, 140, 141, 150-152
Lyons, William 11, 14, 24, 34, 36, 58, 62, 64, 65, 81, 89

M

Mancetter 8
Marque 6, 126, 136, 165
Melksham 41
Mercedes 11, 56-58, 86, 105
Meteor 41-44, 46-48
Mini 143-146, 153
MIRA 10, 44, 83, 84, 99, 136
MK I 54, 72, 76, 78, 83, 101, 102, 106-108, 117, 120-122, 131, 160, 165, 175
MK II 54, 59, 83, 101, 110, 111, 120, 128, 142, 145, 165, 174, 176, 178, 181
MK V 16, 17, 26, 35, 36, 173, 174
MK VII 4, 22, 24, 32, 33, 37, 52, 69, 80, 91, 103

MK VIII 69, 74, 165
MK X 127, 129
Mosquito 46, 47, 49
Moss, Stirling 47, 58, 112, 133
Mulliner 54, 166

N

Nicholson, Bill 79, 81, 88, 89, 130, 134
Nöcker, Peter 95, 147-149
Northallerton 42, 44, 49
Nuneaton 17, 28, 33, 45, 118, 119, 143, 173

P

Pilot 9, 10, 37, 42, 47, 87, 104
Pressed Steel 52, 55, 58
Production Line 14, 15, 19, 21, 24-29, 31, 35, 36, 42, 52-56, 58, 59, 61, 62, 67, 76, 77, 82, 109, 110, 121, 124, 135, 142, 145, 151, 161, 164, 166, 168, 170

R

Radford 15, 112, 164
Radio 9, 41, 57, 105
Rainbow, Frank 80, 82, 86, 137
Reynolds 11
Riley 11, 54, 91
Robinson, Geoffrey 166-169
Rolls-Royce 2, 8, 11
Royal Air Force 4, 9, 10, 38-40, 42, 45-50, 83, 87, 109, 120, 151, 173, 176

S

Salvadori, Roy 133, 134, 137, 141, 142
Sayer, Malcolm 105, 132, 147, 149, 151, 170
Silverstone 22, 27, 111, 112, 115, 116, 134, 145, 160, 180
Standard Triumph 11, 54, 91

Sunbeam 11, 12, 89
Sutton, Ron 'Soapy' 20, 62, 80, 82, 103
Swallow Company 14
Swallow Road 6, 11, 13, 14, 17, 20, 28, 34, 42, 52, 82, 166

T

Tattersall, Burt 76, 77, 82, 83, 146, 150, 152
Turner, 'Wally' 38, 128, 163

V

Vicki 24, 30-34

W

Weaver, Phil 80, 82, 105, 106, 116, 132, 136, 138, 147
White, Derek 109, 131, 137, 149-152
White & Poppe 14, 15
Whyte, Andrew 7, 85
Wolseley 49, 50
Woods, Barrie 81, 113, 116, 135, 174
Woods, Stanley 90, 118-120, 178
Woodwork 10, 18, 22, 66, 176
Workshop 10, 22, 80, 116, 130, 143, 160
Wright, Eric 147

X

XJ6 138, 150, 152, 164, 170, 178
XJ12 127, 128, 152, 170
XJ13 82, 149-152, 155
XJC 5, 169, 170, 178-180
XJS 5, 69, 169-171, 181
XK120 4, 6, 20-27, 30, 33-38, 56, 68, 69, 96, 99, 103, 120, 148, 149, 174
XK140 52, 67-69, 74, 87
XK150 91, 104-106, 128, 134
XKSS 4, 70-72, 94, 97-98, 100, 112, 114, 177, 179-180